Eukaryotic
Gene Expression

GWUMC Department of Biochemistry Annual Spring Symposia

Series Editors:
Allan L. Goldstein, Ajit Kumar, and George V. Vahouny
The George Washington University Medical Center

DIETARY FIBER IN HEALTH AND DISEASE
Edited by George V. Vahouny and David Kritchevsky

EUKARYOTIC GENE EXPRESSION
Edited by Ajit Kumar

Eukaryotic Gene Expression

Edited by
Ajit Kumar
The George Washington University Medical Center
Washington, D.C.

Plenum Press   •   New York and London

Library of Congress Cataloging in Publication Data

Main entry under title:

Eukaryotic gene expression.

(GWUMC Department of Biochemistry annual spring symposia)
Bibliography: p.
Includes index.
1. Gene expression. I. Kumar, Ajit. II. George Washington University. Medical
Center. Dept. of Biochemistry. III. Series. [DNLM: 1. Cytology—Congresses. 2.
Genes—Congresses. 3. Gene expression regulation—Congresses. QH 450 E856
1982]
QH450.E93 1984   574.87'322   83-24721
ISBN 0-306-41532-1

Contributors

W. F. ANDERSON
Laboratory of Molecular Hematology
National Heart, Lung, and Blood Institute
National Institutes of Health
Bethesda, Maryland 20205

A. BAUR
Clinical Hematology Branch
National Heart, Lung, and Blood Institute
National Institutes of Health
Bethesda, Maryland 20205

P. BERG
Laboratory of Molecular Hematology
National Heart, Lung, and Blood Institute
National Institutes of Health
Bethesda, Maryland 20205

S. BERNSTEIN
Laboratory of Molecular Hematology
National Heart, Lung, and Blood Institute
National Institutes of Health
Bethesda, Maryland 20205

PAUL A. BIRO
Department of Human Genetics
Yale University School of Medicine
New Haven, Connecticut 06510
Present address:
Biological Laboratories
Harvard University
Cambridge, Massachusetts 02138

WILLIAM M. BONNER
Laboratory of Molecular Pharmacology
Division of Cancer Treatment
National Cancer Institute
National Institutes of Health
Bethesda, Maryland 20205

IAIN L. CARTWRIGHT
Department of Biology
Washington University
St. Louis, Missouri 63130

BARBARA CHRISTY
Department of Biology
The Johns Hopkins University
Baltimore, Maryland 21218

VAL CIZEWSKI
The Johns Hopkins University
School of Medicine
Baltimore, Maryland 21205

JOHN F. CONNAUGHTON
Department of Biochemistry
The George Washington University School
  of Medicine
Washington, D.C. 20037

NANCY E. COOKE
Endocrine Section, Department of Medicine
and Department of Human Genetics
University of Pennsylvania
Philadelphia, Pennsylvania 19104

HRIDAY K. DAS
Department of Human Genetics
Yale University School of Medicine
New Haven, Connecticut 06510

BENOIT DE CROMBRUGGHE
Laboratory of Molecular Biology
National Cancer Institute
National Institutes of Health
Bethesda, Maryland 20205

HENRI DE GREVE
Laboratorium voor Genetische Virologie
Vrije Universiteit Brussel
B-1640 Sint-Genesius-Rode, Belgium

ANN DEPICKER
Laboratorium voor Genetica
Rijksuniversiteit Gent
B-9000 Gent, Belgium

J. DiPIETRO
Laboratory of Molecular Hematology
National Heart, Lung, and Blood Institute
National Institutes of Health
Bethesda, Maryland 20205

SARAH C. R. ELGIN
Department of Biology
Washington University
St. Louis, Missouri 63130

GERHARD FLEISCHMANN
Department of Biology
Washington University
St. Louis, Missouri 63130

JON W. GORDON
Department of Biology
Yale University
New Haven, Connecticut 06511
Present address:
Department of Obstetrics and Gynecology
Mount Sinai School of Medicine
New York, New York 10029

CORNELIA GORMAN
Laboratory of Molecular Biology
National Cancer Institute
National Institutes of Health
Bethesda, Maryland 20205

PETER GRUSS
Laboratory of Molecular Virology
National Cancer Institute
National Institutes of Health
Bethesda, Maryland 20205

JEAN-PIERRE HERNALSTEENS
Laboratorium voor Genetische Virologie
Vrije Universiteit Brussel
B-1640 Sint-Genesius-Rode, Belgium

LUIS HERRERA-ESTRELLA
Laboratorium voor Genetica
Rijksuniversiteit Gent
B-9000 Gent, Belgium

MARCELLE HOLSTERS
Laboratorium voor Genetica
Rijksuniversiteit Gent
B-9000 Gent, Belgium

BRUCE HOWARD
Laboratory of Molecular Biology
National Cancer Institute
National Institutes of Health
Bethesda, Maryland 20205

R. K. HUMPHRIES
Clinical Hematology Branch
National Heart, Lung, and Blood Institute
National Institutes of Health
Bethesda, Maryland 20205

HENK JOOS
Laboratorium voor Genetica
Rijksuniversiteit Gent
B-9000 Gent, Belgium

MICHAEL A. KEENE
Department of Biology
Washington University
St. Louis, Missouri 63130

GEORGE KHOURY
Laboratory of Molecular Virology
National Cancer Institute
National Institutes of Health
Bethesda, Maryland 20205

AJIT KUMAR
Department of Biochemistry
The George Washington University School
   of Medicine
Washington, D.C. 20037

LAIMOMIS LAIMONS
Laboratory of Molecular Virology
National Cancer Institute
National Institutes of Health
Bethesda, Maryland 20205

RAYMOND E. LOCKARD
Department of Biochemistry
The George Washington University School
  of Medicine
Washington, D.C. 20037

KY LOWENHAUPT
Department of Biology
Washington University
St. Louis, Missouri 63130

GLENN T. MERLINO
Laboratory of Molecular Biology
National Cancer Institute
National Institutes of Health
Bethesda, Maryland 20205

KATHRYN G. MILLER
The Johns Hopkins University
School of Medicine
Baltimore, Maryland 21205

MARC VAN MONTAGU
Laboratorium voor Genetica
Rijksuniversiteit Gent
B-9000 Gent, Belgium
and Laboratorium voor Genetische Virologie
Vrije Universiteit Brussel
B-1640 Sint-Genesius-Rode, Belgium

A. NIENHUIS
Clinical Hematology Branch
National Heart, Lung, and Blood Institute
National Institutes of Health
Bethesda, Maryland 20205

JULIAN PAN
Department of Human Genetics
Yale University School of Medicine
New Haven, Connecticut 06510

IRA PASTAN
Laboratory of Molecular Biology
National Cancer Institute
National Institutes of Health
Bethesda, Maryland 20205

DENNIS PEREIRA
Department of Human Genetics
Yale University School of Medicine
New Haven, Connecticut 06510
Present address:
Microbiological Genetics
Pfizer Central Research
Groton, Connecticut 06340

VEMURI B. REDDY
Department of Human Genetics
Yale University School of Medicine
New Haven, Connecticut 06510
Present address:
Integrated Genetics
Framingham, Massachusetts 01701

RONALD REEDER
Hutchinson Cancer Research Center
Seattle, Washington 98104

JUDITH ROAN
Hutchinson Cancer Research Center
Seattle, Washington 98104

FRANK H. RUDDLE
Department of Biology
Yale University
New Haven, Connecticut 06511

GEORGE SCANGOS
Department of Biology
The Johns Hopkins University
Baltimore, Maryland 21218

JEFF SCHELL
Laboratorium voor Genetica
Rijksuniversiteit Gent
B-9000 Gent, Belgium
and Max-Planck-Institut für Züchtungsforschung
D-5000 Cologne 30, Federal Republic of
  Germany

JO SCHRÖDER
Max-Planck-Institut für Züchtungsforschung
D-5000 Cologne 30, Federal Republic of
  Germany

BARBARA SOLLNER-WEBB
The Johns Hopkins University
School of Medicine
Baltimore, Maryland 21205

ASHWANI K. SOOD
Department of Human Genetics
Yale University School of Medicine
New Haven, Connecticut 06510

ELIZABETH STEINER
Department of Biology
Washington University
St. Louis, Missouri 63130

JAYA SIVASWAMI TYAGI
Laboratory of Molecular Biology
National Cancer Institute
National Institutes of Health
Bethesda, Maryland 20205

SHERMAN M. WEISSMAN
Department of Human Genetics
Yale University School of Medicine
New Haven, Connecticut 06510

RON WIDES
The Johns Hopkins University
School of Medicine
Baltimore, Maryland 21205

JoANNE KAYE WILKINSON
The Johns Hopkins University
School of Medicine
Baltimore, Maryland 21205

LOTHAR WILLMITZER
Max-Planck-Institut für Züchtungsforschung
D-5000 Cologne 30, Federal Republic of
    Germany

ROY S. WU
Biotech Research Laboratories, Inc.
Rockville, Maryland 20850
and Laboratory of Molecular Pharmacology
Division of Cancer Treatment
National Cancer Institute
National Institutes of Health
Bethesda, Maryland 20205

PATRICIA ZAMBRYSKI
Laboratorium voor Genetica
Rijksuniversiteit Gent
B-9000 Gent, Belgium

Preface

The recent surge of interest in recombinant DNA research is understandable considering that biologists from all disciplines, using recently developed molecular techniques, can now study with great precision the structure and regulation of specific genes. As a discipline, molecular biology is no longer a mere subspeciality of biology or biochemistry: it is the *new* biology. Current approaches to the outstanding problems in virtually all the traditional disciplines in biology are now being explored using the recombinant DNA technology. In this atmosphere of rapid progress, the role of information exchange and swift publication becomes quite crucial. Consequently, there has been an equally rapid proliferation of symposia volumes and review articles, apart from the explosion in popular science magazines and news media, which are always ready to simplify and sensationalize the implications of recent discoveries, often before the scientific community has had the opportunity to fully scrutinize the developments. Since many of the recent findings in this field have practical implications, quite often the symposia in molecular biology are sponsored by private industry and are of specialized interest and in any case quite expensive for students to participate in. Given that George Washington University is a teaching institution, our aim in sponsoring these Annual Spring Symposia is to provide, at cost, a forum for students and experts to discuss the latest developments in selected areas of great significance in biology. Additionally, since the University is located in Washington, D.C., there is ample opportunity to draw on the input of policy makers and political leadership, which significantly influences the support of biological research.

Eukaryotic gene expression was a logical choice as a topic, since much of the scientific progress in this field directly or indirectly influences the course of development in other areas of biomedical research. In selecting the topics to be included in this volume, my primary concern was to avoid a narrowly specialized set of articles that are quite often overreviewed and instead include topics that are currently of interest, including promising new technical approaches. The chapters in this volume represent three main areas of interest. In the first group, the emphasis is on selected approaches to the organization of genetic material that have proved to be most promising—for example, studying the organization of active sites using the nuclease sensitivity of specific genes in chromatin (Chapter 1), the role of methylated regions in DNA in controlling gene expression such as that of the thymidine kinase gene (Chapter 2), and the significance of histone types synthesized in the regulation of gene expression during the cell cycle (Chapter 3). The structure and evolutionary divergence of an inducible gene, the prolactin gene, is emphasized in Chapter 4.

A second series of chapters concentrates on the expression of specific genes, in each case emphasizing an important technical approach. In Chapter 5, the structural analysis of histocompatibility class I and class II genes cloned by a sensitive method utilizing synthetic oligonucleotides is described by Sood and colleagues. The next two chapters describe gene-transfer experiments in which the primary aim is to study the regulation of their expression during the development of the intact organism. Chapter 6, by Gordon and Ruddle, discusses the general rationale for producing transgenic mice with specific reference to the expression of the herpes thymidine kinase gene and the human leukocyte interferon gene. Chapter 7 focuses on the expression of globin genes in the transgenic mouse.

The final series of chapters is concerned primarily with *in vivo* and *in vitro* transcriptional studies of specific genes. In Chapter 8, Gorman and colleagues describe a most sensitive and unique assay system in which the expression of the gene for chloromphenicol acetyltransferase is utilized to study the eukaryotic transcriptional enhancer elements. Chapter 9, by Schell and colleagues, discusses a very promising approach utilizing the Ti plasmid vector system for the regulation of plant genes. The remaining chapters focus on the organization and the transcriptional control elements of chicken collagen gene (Chapter 10) and the ribosomal genes of mouse and frog (Chapter 11) and rabbit (Chapter 12).

In all, these articles represent important fields of current research in molecular biology, and it is hoped that *Eukaryotic Gene Expression* will be a useful reference volume for students and experts alike.

Ajit Kumar

Washington, D.C.

Contents

Chapter 8

A Novel System Using the Expression of Chloramphenicol Acetyltransferase in Eukaryotic Cells Allows the Quantitative Study of Promoter Elements............................... 129

Cornelia Gorman, Laimomis Laimons, Glenn T. Merlino, Peter Gruss, George Khoury, and Bruce Howard

Chapter 9

Ti Plasmids as Gene Vectors for Plants 141

Jeff Schell, Marc Van Montagu, Marcelle Holsters, Patricia Zambryski, Henk Joos, Luis Herrera-Estrella, Ann Depicker, Jean-Pierre Hernalsteens, Henri De Greve, Lothar Willmitzer, and Jo Schröder

Chapter 10

Activity of a Chick Collagen Gene in Heterologous and Homologous Cell-Free Extracts

*Glenn T. Merlino, Jaya Sivaswami Tyagi,
Benoit de Crombrugghe, and Ira Pastan*

Chapter 11

Transcription of Ribosomal RNA Genes in Mouse and Frog

*Barbara Sollner-Webb, JoAnne Kaye Wilkinson,
Kathryn G. Miller, Ron Wides, Val Cizewski, Ronald Reeder,
and Judith Roan*

Some Observations on DNA Structure and Chromatin Organization at Specific Loci in *Drosophila melanogaster*

MICHAEL A. KEENE, IAIN L. CARTWRIGHT, GERHARD FLEISCHMANN, KY LOWENHAUPT, ELIZABETH STEINER, and SARAH C. R. ELGIN

1. INTRODUCTION

Within the eukaryotic nucleus, the DNA is packaged in a complex fashion by association with histones and other chromosomal proteins. One may suggest *a priori* that differential protein packaging of coding sequences at the broad level of the chromomere, or in the specific vicinity of a gene, or both, might be an important determinant in the selective expression of these sequences. Our goals have been to map features of chromatin structure relative to known functional sequences, to establish the presence of alternative patterns of structure during development, and to look for alterations in structure that might occur as part of the process of gene induction and repression. To this end, we have recently conducted a series of studies utilizing several different

MICHAEL A. KEENE, IAIN L. CARTWRIGHT, GERHARD FLEISCHMANN, KY LOWENHAUPT, ELIZABETH STEINER, and SARAH C. R. ELGIN • Department of Biology, Washington University, St. Louis, Missouri 63130.

DNA-cleavage reagents to examine the patterns of DNA–protein interaction at a number of *Drosophila* genes. Concurrent studies using immunofluorescent staining of polytene chromosomes have identified several presumptive structural nonhistone chromosomal proteins, including some the distribution pattern of which indicates a preferential association with loci that are to be expressed at some point in the development of the salivary gland cells of *Drosophila*. We anticipate that the synthesis of this information may ultimately lead to a better understanding of the process of gene activation and hence provide insights into the regulation of this event during development. For a more thorough review of many of the issues raised herein, see Cartwright *et al.* (1982).

2. DEOXYRIBONUCLEASE-I-HYPERSENSITIVE SITES

Initial studies using deoxyribonuclease (DNase) I to digest chromatin suggested that this enzyme did not recognize any particular structural features, although a more rapid digestion of active loci relative to the genome as a whole was observed (Weintraub and Groudine, 1976). However, an experiment that allowed observation of the initial cleavage event for a given locus indicated an interesting specificity. Aliquots of nuclei isolated from *Drosophila* embryos were briefly digested with increasing amounts of DNase I, and the DNA was then purified and separated on agarose gels. A uniform smear of DNA fragments of decreasing average molecular weight was observed on staining with ethidium bromide. However, if the DNA fragments were then transferred to nitrocellulose by the Southern (1975) blotting technique and visualized by autoradiography using a specific radioactive DNA probe, a reproducible pattern of discrete bands was visualized (Wu *et al.*, 1979a). These bands apparently result from cleavage of the chromatin at particular sites hypersensitive to DNase I. It has been possible to unambiguously determine the position of such hypersensitive sites with respect to the restriction map of a particular region by utilizing an indirect end-labeling procedure as illustrated in Fig. 1 (Wu, 1980; Nedospasov and Georgiev, 1980). In these experiments, purified DNA from nuclear digests is cleaved completely with an appropriate restriction enzyme; *Bam* HI is used in the example shown. The DNA fragments are size-separated on an agarose gel, a Southern blot is prepared, and the filter is hybridized with a recombinant DNA fragment abutting the restriction site of interest, in this case the *Bam–Sal* fragment at the left-hand end. Only fragments to the right of this site will be visualized, as shown by the lines below the map. The largest is the parental *(Bam–Bam)*

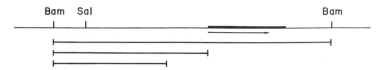

FIGURE 1. Schematic representation of a typical indirect end-labeling experiment. The lower lines represent final digestion products, illustrating how fragment lengths are interpreted to determine preferential cleavage positions. The heavy horizontal bar denotes a transcribed region, and the arrow indicates its direction of transcription.

restriction fragment. The smaller fragments in general will have the *Bam*HI restriction site bordering the probe sequence as one end point and a DNase-I-hypersensitive site as the other. The size of a fragment gives the position of the DNase I site relative to the restriction site. In practice, a region of interest is generally mapped from both ends, to increase accuracy and minimize the chance of artifacts (such as would occur if a DNase I site fell within the probe region).

The set of small heat-shock genes encoded at locus *67B* is ideally suited to such analysis. The four heat-shock genes are encoded within an entirely unique 12-kilobase (kb) region that has been cloned and extensively mapped. Suitable restriction sites and subcloned fragments are available to allow mapping from both ends of the region. An example of the type of data we obtained in a study of this locus is shown in Fig. 2. Four striking pairs of DNase-I-hypersensitive sites were found within this region, and in each case the sites were found to fall at or near the 5' end of one of the four small heat-shock genes, although the embryos used in these experiments had not been subjected to heat-shock conditions. In each pair, a major and a minor site were present. The major site was proximal to the 5' end of the gene, and the minor one was located between 200 and 300 base pairs (bp) upstream (Keene *et al.*, 1981). No other strong preferential cleavage sites were found within this region, but a pair of minor hypersensitive sites found between *hsp 26* and *hsp 23* have now been shown to lie at the 5' end of a developmentally regulated transcript designated *R* (Sirotkin and Davidson, 1982).

A sampling of results obtained by our laboratory and several others is given in Fig. 3. As indicated, such DNase-I-hypersensitive sites have been found at or near the 5' end of a number of different types of genes. These include genes that are generally inactive but rapidly inducible [heat-shock genes (Wu, 1980; Keene *et al.*, 1981)], genes expressed constitutively [ribosomal protein 49 (Wong *et al.*, 1981), alcohol dehydrogenase (ADC-1) (Sledziewski and Young, 1982)], genes regulated by the cell-growth cycle

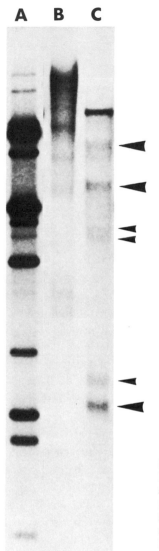

FIGURE 2. A Southern blot containing samples from DNase-I-digested *Drosophila* embryo nuclei. Lanes: (A) size markers; (B) DNA from nuclei digested with DNase I; (C) same DNA purified and digested with *Bam*HI. The blot was hybridized with probe 88.3, a 2.2-kb fragment abutting the *Bam*HI site downstream from *hsp 26*. (◀) Positions of DNase-I-hypersensitive sites of chromatin in locus *67B1*.

[histones (Samal *et al.*, 1981)], and genes expressed in a tissue-specific fashion [duck globin genes (Stalder *et al.*, 1980; Weintraub *et al.*, 1981), mouse globin gene (Sheffery *et al.*, 1982), preproinsulin (Wu and Gilbert, 1981)]. The data come from studies in yeast, *Drosophila*, chick, mouse, and rat. If the gene is one expressed only in certain tissues, one generally observes

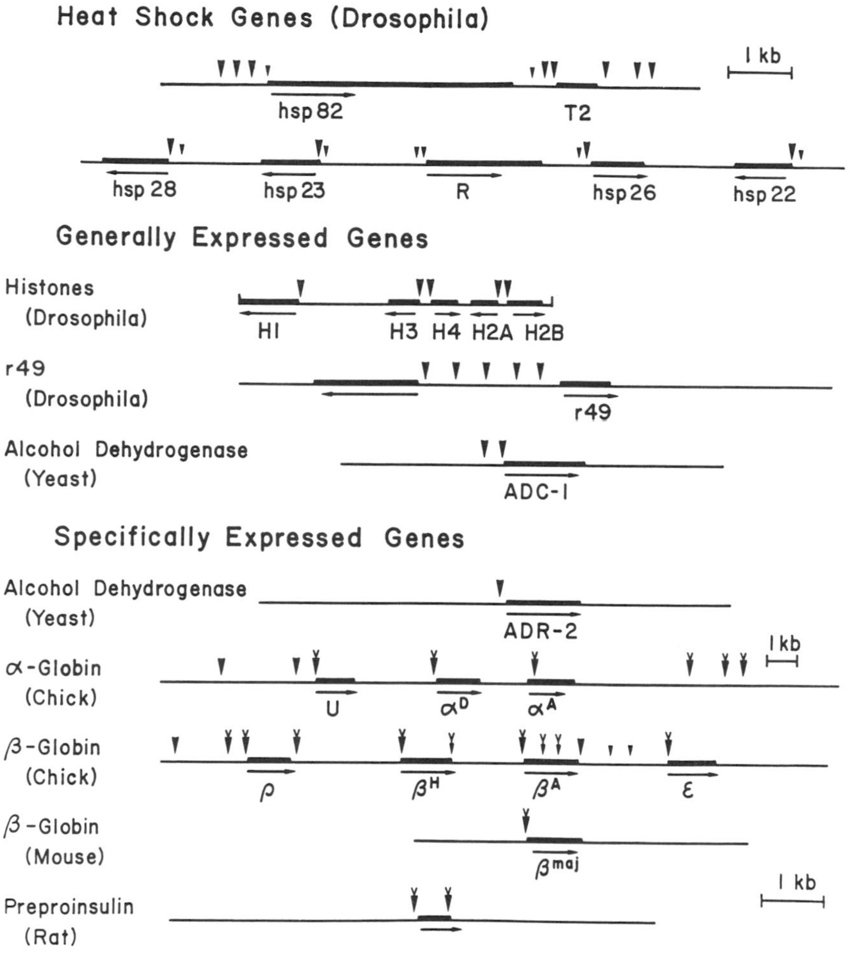

Heat Shock Genes (Drosophila)

FIGURE 3. Summary of some representative DNase-I-hypersensitive sites mapping near regions of transcription. Transcribed regions are shown as heavy lines, and DNase-I-hypersensitive sites are indicated by vertical arrows. Large arrows indicate prominent sites, while smaller arrows represent less preferred sites. DNase-I-hypersensitive sites that have been found only in certain tissues are indicated by fledged arrows.

DNase-I-hypersensitive sites at the 5′ end of the gene only in those tissues (see citations given above). Such DNase-I-hypersensitive sites are also observed near the start sites for transcription of the ribosomal RNA genes of *Tetrahymena* (Borchsenius *et al.*, 1981; Palen *et al.*, 1982). In a few cases, investigators have reported that they cannot detect such a DNase-I-hypersensitive site at the 5′ end of an active gene (e.g., Kuo et al., 1979); the explanation is unknown.

These findings as a whole suggest that a particular chromatin construct (recognized as a DNase-I-hypersensitive site) at or near the 5′ end of many genes is necessary, but not sufficient, for transcription by RNA polymerase II (or I) in vivo. A recent study of the Sgs4 locus of Drosophila melanogaster Oregon R is consistent with this hypothesis. Shermoen and Beckendorf (1982) have observed a cluster of five DNase-I-hypersensitive sites near the 5′ end of this gene that are present in chromatin from salivary glands (where the gene is expressed) but not in chromatin from early embryos (where it is not expressed). A particular strain, BER-1, shows no expression of this locus in any tissue or stage tested (Muskavitch and Hogness, 1980). Comparison of the genomic DNA from D. melanogaster Oregon R and D. melanogaster BER-1 indicates an upstream deletion of approximately 100 bp, which leaves the TATA box intact but eliminates DNA sequences encompassing the major DNase-I-hypersensitive site at −408 (Muskavitch and Hogness, 1982). Although other nearby DNA sequences are preserved, no DNase-I-hypersensitive sites are observed upstream from the glue gene in BER-1 salivary gland chromatin—indicating not only that the major site has been deleted, but also that a critical switch in chromatin structure has not occurred (Shermoen and Beckendorf, 1982). The results suggest that certain sequences 5′ to the gene are necessary for switching the region into an "active" chromatin configuration, but further genetic analysis will be necessary to test this proposition. The topic of DNase-I-hypersensitive sites in chromatin has recently been reviewed (Elgin, 1981; Lowenhaupt et al., 1983a).

The data shown in Fig. 3 suggest that establishment of DNase-I-hypersensitive sites may be a determinative event during differentiation. To examine this question in the most general sense, we have looked at the chromatin structure of a number of loci in preblastoderm Drosophila embryos. These embryos are syncitia; their nuclei display rapid replication, but little or no transcription. When the chromatin structure in the vicinity of the four small heat-shock genes at locus 67B was analyzed, it was found that the pattern of DNase-I-hypersensitive sites was indistinguishable from that obtained when older embryos (postblastula) were used. Similarly, the pattern of hypersensitive sites at the ribosomal protein 49 gene has already been established at this early developmental time (Lowenhaupt et al., 1983b). Results with the heat-shock genes had previously suggested that DNase-I-hypersensitive sites could be found for a given locus prior to transcription; the results with preblastula embryos support and extend this observation. Weintraub et al. (1982) have recently obtained evidence that the formation of DNase-I-hypersensitive sites at globin genes is a relatively late event in the development of chick red blood cells, but clearly precedes transcription. Additional events are required to initiate transcription. Such changes might include a reduced compaction of a large chromosomal domain as well as the interaction of specific effector molecules with their target genes.

What is the structural nature of the DNase-I-hypersensitive site? It is likely that the hypersensitivity results from a relative accessibility of a small region of the chromatin to nucleases. Generally, such "sites" are estimated to involve 40–400 bp of DNA, although it has also been reported that a 1.5-kb region 5' to the ovalbumin gene is DNase-I-sensitive in chick oviduct chromatin (Bellard *et al.*, 1982). In the few cases examined in detail, evidence has been obtained indicating that the hypersensitive site is a nucleosome-free region (Saragosti *et al.*, 1980; McGhee *et al.*, 1981). Whether or not this is a consequence of the association of particular nonhistone chromosomal (NHC) proteins, a reflection of the particular DNA structure involved, or both, is not yet established. Larsen and Weintraub (1982) have reported that there are sequences 5' to the chick globin genes, at or near the DNase-I-hypersensitive sites found in active globin chromatin, that are sensitive to S1 nuclease in chromatin and in supercoiled (but not in linear) purified recombinant DNA plasmid. We have observed one major and one minor S1 site in the supercoiled DNA covering genes *hsp 28* through *hsp 26* at *67B* (Selleck *et al.*, 1984). The data suggest that DNA sequences at these sites can adopt a different conformation dependent on supercoiling of the DNA; it seems possible that such sequences could act as switches within the genome.

3. DNA STRUCTURAL PATTERNS AND NUCLEOSOME DISTRIBUTION

3.1. Micrococcal Nuclease

The discovery of DNase-I-hypersensitive sites in chromatin, specifically positioned in the vicinity of the 5' ends of numerous genes, suggested that the nucleosomes in these regions might occupy defined positions. It appeared that an application of the indirect end-labeling protocol, using micrococcal nuclease as the cleavage reagent, would allow one to map the positions of nucleosomes. If nucleosome linkers, which are preferentially cleaved, were regularly positioned with respect to the DNA sequence, then a reproducible set of discrete bands would appear at nucleosomal intervals in such an experiment. If, however, nucleosomes were randomly positioned with respect to the DNA sequence, then a smear of DNA fragments would appear. While a regular array of bands spaced at nucleosomal intervals was immediately detected in several places within locus *67B,* it was also immediately apparent that the interpretation of such data would not be simple. When purified high-molecular-weight *Drosophila* DNA was subjected to an identical procedure, the majority of bands seen in the nuclear digestion pattern were reproduced in the naked DNA pattern (see Fig. 4) (Keene and Elgin, 1981). Since such results could conceivably arise due to modification or cleavage of the DNA in intact nuclei, analogous experiments were performed using intact super-coiled recombinant plasmids as substrate. Again, the bulk of the discrete banding pattern was reproduced in the plasmid DNA. Plasmids grown in dam$^+$ and dam$^-$ *Escherichia coli* gave identical results, indicating no effect due to methylation of the DNA (Elgin *et al.,* 1983). Some minor preferential cleavage sites can be visualized better in the patterns derived using plasmid (rather than genomic) DNA, but in all cases the pattern of prominent cleavage sites is the same, with major cutting sites at roughly 200-bp intervals. The cleavage pattern clearly reflects an inherent property of the DNA itself that is recognized by micrococcal nuclease (Keene and Elgin, 1981).

Despite this observation, chromatin-specific features can frequently be observed near the 5' end of genes using micrococcal nuclease. In some instances, a preferential cleavage site is present only in intact chromatin, while in others a site present in naked DNA is not observed in nuclei. In all cases, these specific features overlap the DNase-I-hypersensitive sites at or near the 5' ends of the genes in question. Current evidence favors the view that this particular region is nucleosome-free. It had seemed reasonable to suppose that a DNase-I-hypersensitive site might represent a boundary for nucleosome placement, thereby creating local order in the nucleosome distribution. The data are compatible with this possibility, in that the similarity of DNA and chromatin patterns may indicate that micrococcal-nuclease-sensitive sites of DNA tend to fall in linker regions. However, most results do not provide conclusive evidence on this point. In a few instances, it has been possible to successfully map some specifically positioned nucleosomes (Bryan *et al.,* 1981; Bloom and Carbon, 1982). Similarly, while it remains possible that a long-range ordering of the nucleosomal array with respect to the DNA sequence exists, it is clear that this cannot be unequivocably established relying solely on the methodology outlined above.

The nonrandom distribution of prominent micrococcal-nuclease cleavage sites on purified DNA has become a focus of interest in itself. While arrays

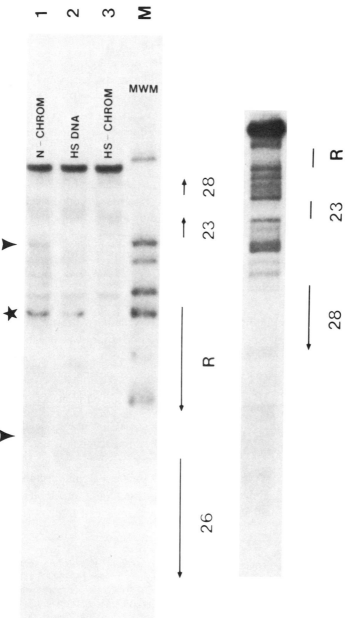

FIGURE 4. Digestion of locus *67B1* by micrococcal nuclease. *Top*—Lanes: (1) DNA from normal 6- to 18-hr embryos, nuclei digested with 120 U/ml micrococcal nuclease; (2) DNA from 6- to 18-hr embryos that were heat-shocked at 36°C for 20 min, purified DNA digested with 6 U/ml nuclease; (3) DNA from heat-shocked embryos, nuclei digested with 120 U/ml nuclease; (M) molecular-weight markers. The DNA in samples 1–3 was restricted completely with *Bam*HI. After electrophoresis through a 0.9% agarose gel, a Southern blot was prepared and hybridized with nick-translated 88.3 as probe. *Bottom*: DNA from plasmid 88B13 digested with 0.05 U/ml micrococcal nuclease, purified, and digested completely with *Bam*HI. The Southern blot of the gel was hybridized with a nick-translated *Bam*HI–*Hind*III fragment from the left-hand end of 88B13.

of regularly spaced cleavage sites have been seen over distances of several kilobases, we observed that such organization was limited to nontranscribed regions. Each of the four small heat-shock genes at *67B* was devoid of prominent cleavage sites and appeared as a gap in autoradiographs from such experiments. In addition, a large gap in the middle of the heat-shock gene cluster predicted the size and location of the *R* gene mentioned in Section 2 (see Fig. 4). The pattern of organization exhibited here has now been shown to hold for several other developmentally regulated genes at *67B,* as well as for transcripts located at the ribosomal protein 49 gene and actin gene clusters (Keene and Elgin, 1984). At this point, it appears likely that this is a general phenomenon in *Drosophila,* when the prokaryotic plasmid pBR322 is examined, a totally different pattern of cleavage is observed. Prominent micrococcal-nuclease cleavage sites are distributed throughout the transcribed, as well as the nontranscribed, regions of the plasmid, and the spacing is much less regular, with an aveage of 165 bp (Keene and Elgin, 1981). A similar contrast between patterns in eukaryotic (yeast) and prokaryotic (pBR322) DNA has been noted by Bloom and Carbon (1982).

It has been suggested that CTA or CATA or both may be the consensus sequences for preferential cutting by micrococcal nuclease (Hörz and Altenberger, 1981). However, in an analysis of the *5S* gene at the sequence level, we have observed that the rate of cleavage of these sequences varies considerably and that many prominent cleavage points bear no obvious relationship to either these sequences or each other (Fig. 5) (Cartwright and Elgin, 1982). In contrast to the situation with S1, the pattern does not vary between linear and supercoiled DNA molecules. The results suggest that the pattern of cutting by micrococcal nuclease is dictated by local structural features that reflect both the immediate sequence and the base composition of the region. Variations in the structure of DNA might facilitate its packaging with histones to form nucleosomes. It is certainly conceivable that structural information read by micrococcal nuclease could also be read by the components of chromatin;

FIGURE 5.   A 6% acrylamide sequencing gel showing preferential cleavage sites for micrococcal nuclease and the 1,10-phenanthroline–cuprous complex [$(OP)_2Cu^i$] in a 3′-end-labeled *DdeI–TaqI* fragment of 5 S DNA from plasmid 12D1 (bases numbered relative to initiation of transcription). (1) Base-specific cleavages according to Maxam and Gilbert, from left to right: G, G + A, C + T, C; (2) $(OP)_2Cu^i$ digestions were for 15 sec and 1, 3, and 8 min at 25°C; (3) micrococcal nuclease digestions were at 0.5 U/ml for 1, 3, 7, and 12 min at 25°C. Bases underlined represent strong micrococcal-nuclease cleavage preference. (Reprinted, with permission, from Cartwright and Elgin, 1982.)

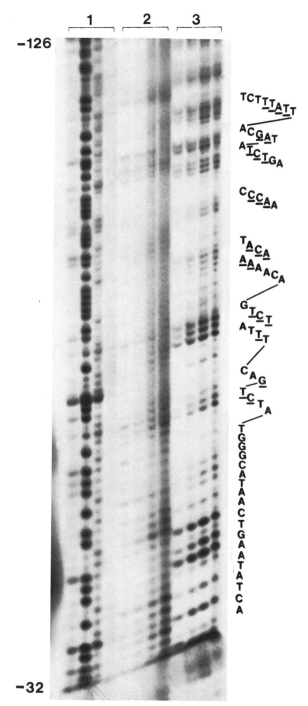

the approximate 200-bp spacing of most cleavage sites in the eukaryotic sequences examined indicates that the chromatin structure could follow this pattern.

3.2. Chemical Cleavage

The indirect end-labeling procedure should allow one to unambiguously resolve the question of nucleosome placement assuming that cleavage reagents that are indifferent to DNA sequence, yet able to recognize the difference between nucleosome cores and linkers, can be found. The putative intercalator 1,10-phenanthroline–cuprous complex was reported to be relatively sequence-neutral in its cutting specificity (Marshall *et al.*, 1981) and was shown to produce a regular oligonucleosomal ladder on incubation with *Drosophila* nuclei (Cartwright and Elgin, 1982). The 1,10-phenanthroline–cuprous complex is an effective reagent for degradation of double-stranded DNA, apparently utilizing intercalation to efficiently localize the heavy-metal ion. The chemical basis of the degradation probably derives from the production (and subsequent reaction with DNA) of the hydroxyl radical, either from dissolved oxygen or from added hydrogen peroxide, mediated by oxidation of the metal ion (Que *et al.*, 1980; Marshall *et al.*, 1981). It is possible to use this reagent under rather different ionic conditions than those required for most nucleases, thus minimizing degradation by endogenous enzymes.

When samples generated using the 1,10-phenanthroline–cuprous complex were treated as outlined above for an indirect end-label experiment, the pattern of prominent cleavages in the vicinity of locus *67B* was found to be virtually indistinguishable from that produced using micrococcal nuclease. Patterns from DNA and chromatin were very similar, although chromatin-specific structures could be detected around the 5' ends of the genes. While some variation in the intensity of specific bands was noted, the overall organization, position, and spacing of prominent cleavages were very similar with the two different reagents (Fig. 6). Note that at the level of resolution of a sequencing gel, it appears that the precise site of action is not identical in all cases (see Fig. 5). However, at the level of resolution of a 1% agarose gel, one sees an overall similarity. The data in hand are not sufficient to determine a consensus sequence for the preferred cutting site of the 1,10-phenanthroline–cuprous complex.

The finding that two agents that achieve DNA cleavage by different mechanisms generate similar patterns, and so presumably are recognizing the

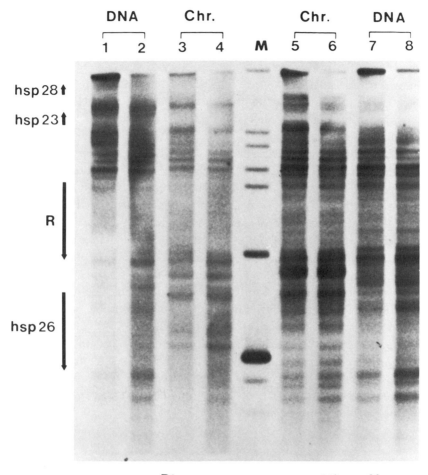

FIGURE 6. Digestions at locus *67B1* in both isolated nuclei and protein-free genomic DNA. Genomic DNA at 100 μg/ml in 10 mM Tris pH 8.0/5 mM NaCl/1 mM EDTA was incubated with $(OP)_2Cu^I$ and 1.67 mM H_2O_2 at 25°C for 1 min (lane 1) and 2 min (lane 2); similarly, nuclei were digested with the complex in 0.4 mM H_2O_2 for 2 min (lane 3) and 4 min (lane 4). Micrococcal-nuclease digestions at 25°C were performed on nuclei for 3 min with 6 U/ml (lane 5) and 12 U/ml (lane 6) of enzyme; similarly, genomic DNA at 500 μg/ml was digested with 12 U/ml (lane 7) and 18 U/ml (lane 8) of enzyme. Samples were subsequently cut to completion with *Bam*HI, loaded on a 0.9% agarose gel, and a Southern blot prepared and probed with nick-translated plasmid 88.3. (M) Molecular-weight markers. (Reprinted, with permission, from Cartwright and Elgin, 1982.)

same feature(s) of these DNA sequences, has suggested that the cleavage pattern reflects some fundamental structural pattern in the DNA; however, we know very little about the nature of the recognition events (Cartwright and Elgin, 1982). It will probably be necessary to generate much more data from experiments of this type to be able to predict the preferred cleavage site(s). The results obtained to date suggest that reagents such as these are providing a map of DNA structure and that this structure is related to the organization of genes.

Recently, we have initiated a series of studies using a new intercalating reagent, methidium propyl-EDTA (MPE). This synthetic molecule, when complexed to ferrous ion, works by a mechanism analogous to that of the 1,10-phenanthroline–cuprous complex in cleaving DNA (Hertzberg and Dervan, 1982). Studies on prokaryotic DNA indicate that the cleavage reaction is relatively sequence-neutral, well suited to the footprinting studies of drug–DNA interactions for which it was designed (vanDyke *et al.*, 1982). We have found that under appropriate conditions, MPE can be used to generate a nucleosome array analogous to that produced using micrococcal nuclease (Cartwright *et al.*, 1983). This reagent differs notably, however, from the two discussed above, in that in some regions of the genome, little specificity is observed in cleavage of purified DNA. In other regions, a discrete pattern, with cutting approximately once per 200 bp, is noted. MPE may therefore prove to be very useful in analyzing chromatin structure in certain regions of the genome, but the range of its applicability is not yet established. Other reagents for this purpose are also being explored.

4. CHROMATIN STRUCTURAL CHANGES ASSOCIATED WITH GENE EXPRESSION

4.1. Conditions That Appear Necessary But Not Sufficient

One of the major goals in the study of chromatin structure is to discern those changes that occur during the process of gene activation and transcription and to address the question of when and how structural parameters are used in the control of gene expression. Results from many laboratories suggest that activation is a multistep process with several structural conditions, each necessary but not sufficient. As discussed in Section 2, the data at hand indicate that a specific chromatin structure at the 5′ end of a gene, generally detected as one or more DNase-I-hypersensitive sites, is one such conditon. Evidence from work in *Drosophila* and in chick indicates that this feature of chromatin structure is required prior to transcription at many loci, rather than being a consequence of transcription. It should of course be noted that we do not as yet understand the functional role of the DNase-I-hypersensitive site, although several possibilities come to mind. We should anticipate that in some instances, these functions could be met by mechanisms other than generation of a structural site of the type we commonly observe in *Drosophila*. For example, Bellard *et al.* (1982) have reported an extended region of 1.5 kb 5' to the active ovalbumin gene sensitive to DNase I rather than a specific site.

As previously discussed, it appears that the generation of a DNase-I-hypersensitive site 5' to the gene is frequently a necessary step, but (usually) not the final step, in allowing initiation of transcription. Consequently, the generation of a cell-specific pattern of DNase-I-hypersensitive sites during cell determination and development could be used as a regulatory mechanism. One might ask, for a given cell type, whether the basic structural transition is reversible or whether it is a way of (normally) channeling events in one direction. Our observations on locus *67B* indicate that in some instances, it can be reversible. Locus *67B* includes both genes expressed at a high level in response to heat shock and genes not expressed (presumably actively re-pressed where necessary) in response to heat shock. It was noted that a site hypersensitive to cleavage by micrococcal nuclease in normal nuclei, genomic DNA, and plasmid DNA was obscured in nuclei from heat-shocked cells (see Fig. 4). Further analysis revealed that a DNase-I-hypersensitive site in the same position was similarly obscured under conditions of heat shock. These sites were shown to be at the 5' end of the developmentally regulated tran-script, *R*, in the *67B* locus (Fig. 4). Similar phenomena have now been seen for at least one other developmentally regulated gene in the *67B* cluster. In an effort to determine the generality of this phenomenon, several other genes, including ribosomal protein 49 and a cytoplasmic actin, have been examined. At these loci, no appreciable changes in 5' sites were seen to occur under heat shock, a condition under which transcription is known to be repressed (Keene and Elgin, 1982). It is well known that under conditions of heat shock, chromomeres containing heat-shock genes puff as their transcriptional activity increases, while other puffs representing normally active loci visibly regress (Ashburner and Bonner, 1979). These results suggest that within an active locus (or gene cluster), expression of individual genes can be regulated by manipulation of the chromatin structure at the 5' ends, but that when the locus or cluster can be regulated as a unit, other mechanisms may be involved.

One of the first reports of chromatin structural changes correlated with gene expression concerned the observation by Weintraub and Groudine (1976) that the chromatin of active globin loci was more sensitive to digestion by DNase I than the bulk chromatin. This sensitivity is observed for genes transcribed at low rates as well (Garel *et al.*, 1977). In terminally differentiated cells, the sensitivity of genes such as globin or ovalbumin persists regardless of the actual level of transcription (Weintraub and Groudine, 1976; Palmiter *et al.*, 1978). This broad sensitivity occurs over a region extending well beyond the region of transcription; in the one case where a complete map has been obtained, O'Malley and colleagues have reported that the active oval-bumin gene lies within a sensitive region that covers 100–120 kb (Lawson *et al.*, 1982). It will be of interest to map such domains in *Drosophila*, where one can ask in particular whether or not there is any correlation with the extent of the chromomere as established from cytological and genetic techniques.

4.2. Disruption of the Nucleosomal Array

Digestion of chromatin with micrococcal nuclease alone quickly revealed a disruption of the normal nucleosomal array concomitant with gene induction. When *Drosophila* tissue-culture cells were heat-shocked, no overall perturbation of the chromatin structure was detected by ethidium bromide staining. However, when the genes encoding the 70,000-dalton heat-shock protein were specifically examined, it was found that the normal oligonucleosomal ladder had almost completely disappeared, a smear of fragments being observed instead, and digestion of the gene was disproportionately rapid (Wu *et al.*, 1979b). By subcloning small fragments, it was possible to demonstrate that this region of disruption extends for approximately 2.5 kb beyond the poly-adenylic acid addition site for one copy of the *hsp 70* gene at *87A* (Keene and Elgin, 1982). It was noted that when heat-shocked cells were permitted to recover at normal growth temperature for 3 hr, the normal nucleosomal organization at this locus was regained (Wu *et al.*, 1979b). During this time, less than 10% of the cell population would be expected to undergo replication. It was also found that terminally differentiated Kc cells could undergo this cycle of disruption and recovery, indicating that this alteration in chromatin structure is reversible in differentiated cells (M. A. Keene and S. C. R. Elgin, unpublished observations). Further experiments are in progress to map the boundaries of perturbation relative to the region of transcription at several loci.

Several other investigators have also reported a perturbation of the oligonucleosome array at genes being actively transcribed (e.g., Bellard *et al.*, 1982; Bloom and Anderson, 1979). There is some indication that the degree of perturbation may be a function of the rate of transcription, perhaps reflecting events surrounding the RNA polymerase–DNA interaction. Whether these changes must precede or follow as a consequence of transcription is not yet established.

4.3. Nonhistone Chromosomal Proteins

It appears reasonable to suppose that the generation and stabilization of alternative chromatin structures will require modification of the interactions of the macromolecules involved and that certain NHC proteins will play key roles in this process. Because we are attempting to understand a structural transition, not an enzymatic reaction, it is difficult to devise pertinent assays. We have chosen to look at the distribution of chromosomal proteins in the polytene chromosomes of *Drosophila* using an immunofluorescence assay (Silver and Elgin, 1976). The details of this method, as well as a discussion of its strengths and weaknesses, have been reviewed elsewhere (Silver and Elgin, 1978). In particular, it should be noted that a positive signal will reflect both the distribution of the protein in question and its relative accessibility to the antibody probe.

The results to date indicate that there are at least two classes of chromosomal proteins preferentially associated with active loci. First, proteins involved in transcription and packaging of RNA, the machinery of that process, are found preferentially associated with those loci being expressed at the time of assay. For example, the results obtained using antibodies against RNA polymerase II to stain chromosomes from a heat-shocked larva are shown in Fig. 7 (Elgin *et al.*, 1978). Prominent staining of the heat-shock loci is observed. Second, certain proteins appear to be preferentially associated not only with these loci, but also with the set of loci that will be (or have been) expressed at some time in this particular tissue (salivary gland) (Fig. 8) (Mayfield *et al.*, 1978). One may suggest that these proteins are a part of the chromatin structure of all loci that can be activated in this terminally differentiated cell. It will be of interest to see whether this distribution pattern can be correlated directly with the tissue-specific pattern of broad DNase I sensitivity or some other measure of chromatin structure.

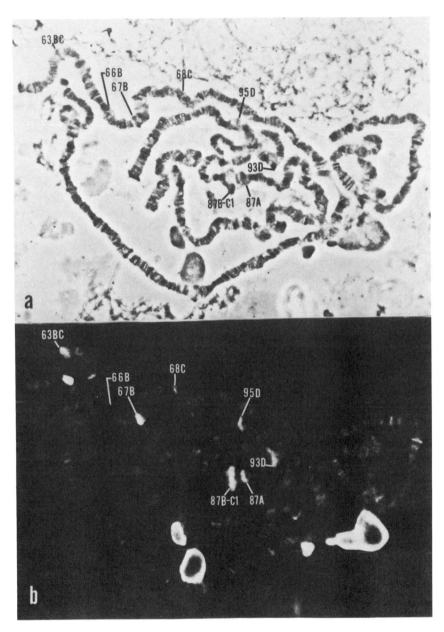

FIGURE 7. Light micrographs of polytene chromosomes from the salivary gland of a heat-shocked third-instar larva, fixed with acetic acid and stained with antibodies against RNA polymerase II. The antiserum was a gift from Arno Greenleaf and E. Bautz. (a) Phase-contrast; (b) fluorescence micrographs. Several prominently staining heat-shock puffs are identified. (Reprinted, with permission, from Elgin *et al.*, 1978.)

5. SUMMARY: DNA STRUCTURE AND CHROMATIN STRUCTURE

We are beginning to develop a functional picture of the eukaryotic genome, not just as a collection of large DNA molecules, but as a complex protein–DNA structure. Gene expression appears to require a set of conditions, each necessary but not sufficient; some apparently reflect the structure of larger units, while others may be necessary for the interaction of RNA polymerase with "promoters," suggesting final events similar to those familiar from work in prokaryotes. The precision of the results achieved (the development of a eukaryotic organism) suggests specificity in chromatin organization. As discussed in Section 3, this has proved to be a difficult subject to explore. Clearly, there are certain sites, such as the DNase-I-hypersensitive sites, that are well-defined, mappable chromatin structures. In addition to such sites at the 5' end of genes, DNase-I-hypersensitive sites have also been found at origins of replication (Palen *et al.*, 1982), sites of DNA rearrangement events (Nasmyth, 1982), and elsewhere, and are also found at other positions with no designated function as yet. Such sites might well serve as boundaries for nucleosome arrays. Boundaries could also be constructed from particular protein–DNA interactions, RNA–DNA hybrids, or DNA sequences such as polydeoxyadenylic acid–polydeoxythymidylic acid that do not readily assume a nucleosome configuration (Dunn and Griffith, 1980; Kunkel and Martinson, 1981). Alternative DNA conformations, such as Z-DNA, are also candidates (Nickol *et al.*, 1982).

Within such boundaries, nucleosomes might assume a statistical distribution or might adopt specific positions along the DNA sequence. While it is well known that most DNA sequences (including those from prokaryotes and synthetic oligonucleotides) can be folded by interaction with a histone core into a nucleosome configuration, recent studies with small defined fragments of DNA indicate preferred positions that are sequence-dependent (Chao *et al.*, 1979, 1980). In one case, a sequence-dependent array of nucleosomes has been detected within a eukaryotic nucleus, around two yeast centromeres (Bloom and Carbon, 1982).

It is of interest to consider the extent to which chromatin structure will be a reflection of DNA structure. Recent studies have shown that the conformation of the DNA double helix, including the structure of the sugar–phosphate backbone, is variable and sequence-dependent, even within the overall B form. This is clear both from recent studies of the B-DNA helical periodicity (Peck and Wang, 1981; Rhodes and Klug, 1981) and from

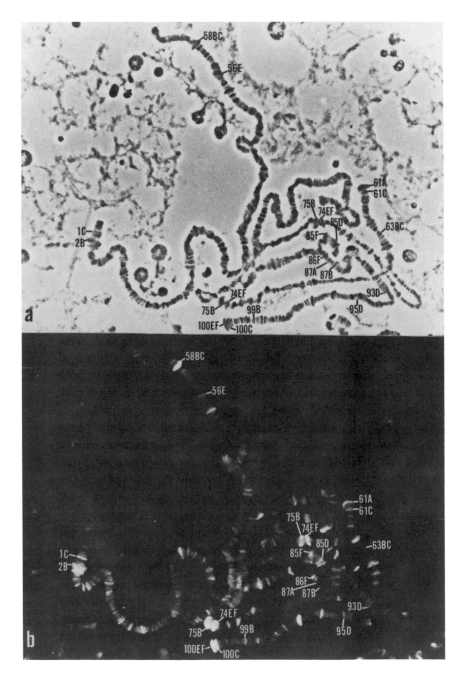

the X-ray crystallographic analysis of the B-DNA dodecamer CGCGAATTCGCG (Dickerson and Drew, 1981a,b). The sequence-dependent conformation is apparently reflected in the sensitivity of the backbone to nucleases, there being a good correlation between the sites of rapid DNase I cutting within this dodecamer and positions of high local helical twist (Lomonossoff *et al.*, 1981). The rate of cleavage at a particular site (and so, presumably, the DNA structure) will reflect not only the immediate site but also the "DNA environment," as illustrated by the fact that the four different *Pst*I sites in pSM1 are cut at very different rates (Armstrong and Bauer, 1982). This effect is not dependent on superhelical twist. It seems possible, then, that one will obtain empirical maps of DNA structure by examining the interaction of DNA with enzymes (such as nucleases) or with other small DNA-binding compounds (such as intercalators). As discussed in Section 3, at locus *67B* of *Drosophila,* initial digestion of purified genomic DNA (or recombinant plasmids) with micrococcal nuclease yields a pattern with prominent cleavage sites tending to occur in the spacer regions and not within the transcribed regions (Keene and Elgin, 1981). Similar cleavage patterns are obtained with the 1,10-phenanthroline–cuprous complex (a presumptive intercalator); again, these sites are spaced at approximately 200-bp intervals and are not dependent on supercoiling (Cartwright and Elgin, 1982). Patterns from chromatin and DNA are remarkably similar in many instances. While the data are not conclusive, they suggest that within certain regions of the genome, one sees patterns of DNA structure that are reflected in patterns of chromatin structure. This is not seen within genes, and probably will not be seen in regions of the genome where information coding is required. Clearly, the DNA pattern can be "overridden" in the determination of chromatin structure, as is apparent at the 5' ends of genes; nonetheless, one may suggest that relatively stable features of DNA structure may be dictating patterns of chromatin structure. If these ideas are correct, it will suggest an additional coding potential for DNA–protein interactions, based not on the sequence of bases *per se* but on the overall structure of the double helix.

FIGURE 8. Chromosomal staining pattern obtained using anti-Band 2 serum. Salivary glands were obtained from a late third-instar larva grown at 25°C and processed through a formaldehyde fixation technique. (a) Phase contrast; (b) fluorescence micrographs. In this squash, chromosome arm 3L is split from Band 64C to the chromocenter. A number of brightly stained, developmentally active loci are labeled, as well as the (normally inactive) heat-shock loci. (Reprinted, with permission, from Mayfield *et al.*, 1978.)

REFERENCES

Armstrong, K., and Bauer, W. R., 1982, Preferential site-dependent cleavage by restriction endonuclease Pst I, *Nucleic Acids Res.* **10**:993–1007.

Ashburner, M., and Bonner, J. J., 1979, The induction of gene activity in *Drosophila* by heat shock, *Cell* **17**:241–254.

Bellard, M., Dretgen, G., Bellard, F., Oudet, P., and Chambon, P., 1982, Disruption of the typical chromatin structure in a 2500 base-pair region at the 5' end of the actively transcribed ovalbumin gene, *Eur. Mol. Biol. Assoc. J.* **1**:223–230.

Bloom, K. S., and Anderson, J. N., 1979, Conformation of ovalbumin and globin genes in chromatin during differential gene expression, *J. Biol. Chem.* **254**:10,532–10,539.

Bloom, K. S., and Carbon, J., 1982, Yeast centromere DNA is a unique and highly ordered structure in chromosomes and small circular minichromosomes, *Cell* **29**:305–317.

Borchsenius, S., Bonven, B., Leer, J. C., and Westergaard, O., 1981, Nuclease-sensitive regions on the extrachromosomal r-chromatin from *Tetrahymena pyriformis*, *Eur. J. Biochem.* **117**:245–250.

Bryan, P. N., Hofstetter, H., and Birnstiel, M. L., 1981, Nucleosome arrangement on rRNA genes of *Xenopus laevis*, *Cell* **27**:459–466.

Cartwright, I. L., and Elgin, S. C. R., 1982, Analysis of chromatin structure and DNA sequence organization: Use of the 1,10-phenanthroline–cuprous complex, *Nucleic Acids Res.* **10**:5835–5853.

Cartwright, I. L., Keene, M. A., Howard, G. C., Abmayr, S. M., Fleischmann, G., Lowenhaupt, K., and Elgin, S. C. R., 1982, Chromatin structure and gene activity: The role of nonhistone chromosomal proteins, *CRC Crit. Rev. Biochem.* **13**:1–86.

Cartwright, I. L., Hertzberg, R. P., Dervan, P. B., and Elgin, S. C. R., 1983, Cleavage of chromatin with (methidiumpropyl-EDTA) iron (II) *Proc. Natl. Acad. Sci. USA* **80**:3213–3217.

Chao, M. V., Gralla, J., and Martinson, H. G., 1979, DNA sequence directs placement of histone cores on restriction fragments during nucleosome formation, *Biochemistry* **18**:1068–1074.

Chao, M. V., Gralla, J. D., and Martinson, H., 1980, lac operator nucleosomes. 1. Repressor binds specifically to operator within the nucleosome core, *Biochemistry* **19**:3254–3260.

Dickerson, R. E., and Drew, H. R., 1981a, Kinematic model for B-DNA, *Proc. Natl. Acad. Sci. U.S.A.* **78**:7318–7322.

Dickerson, R. E., and Drew, H. R., 1981b, Structure of a B-DNA dodecamer. II. Influence of base sequence on helix structure, *J. Mol. Biol.* **149**:761–786.

Dunn, K., and Griffith, J. D., 1980, The presence of RNA in a double helix inhibits its interaction with histone protein, *Nucleic Acids Res.* **8**:555–572.

Elgin, S. C. R., 1981, Minireview: DNAase I-hypersensitive sites of chromatin, *Cell* **27**:413–415.

Elgin, S. C. R., Serunian, L. A., and Silver, L. M., 1978, Distribution patterns of *Drosophila* nonhistone chromosomal proteins, *Cold Spring Harbor Symp. Quant. Biol.* **42**:839–850.

Elgin, S. C. R., Cartwright, I. L., Fleischmann, G., Lowenhaupt, K., and Keene, M. A., 1983, Cleavage reagents as probes of DNA sequence organization and chromatin structure: *Drosophila melanogaster* locus 67B1, *Cold Spring Harbor Symp. Quant. Biol.* **47**:529–538.

Garel, A., Zolan, M., and Axel, R., 1977, Genes transcribed at diverse rates have a similar conformation in chromatin, *Proc. Natl. Acad. Sci. U.S.A.* **74**:4867–4871.

Hertzberg, R. P., and Dervan, P. B., 1982, Cleavage of double helical DNA by (methidium-propyl-EDTA) iron (II), *J. Am. Chem. Soc.* **104**:313–315.

Hörz, W., and Altenberger, W., 1981, Sequence specific cleavage of DNA by micrococcal nuclease, *Nucleic Acids Res.* **9**:2643–2658.

Keene, M. A., and Elgin, S. C. R., 1981, Micrococcal nuclease as a probe of DNA sequence organization and chromatin structure, *Cell* **27**:57–64.

Keene, M. A., and Elgin, S. C. R., 1982, Perturbations of chromatin structure associated with gene expression, in: *Heat Shock: From Bacteria to Man* (M. J. Schlesinger, M. Ashburner, and A. Tissieres, eds.), Cold Spring Harbor Laboratory, New York, pp. 83–90.

Keene, M. A., and Elgin, S. C. R., 1984, Patterns of DNA structural polymorphism and their evolutionary implications (submitted).

Keene, M. A., Corces, V., Lowenhaupt, K., and Elgin, S. C. R., 1981, DNase I hypersensitive sites in *Drosophila* chromatin occur at the 5' ends of regions of transcription, *Proc. Natl. Acad. Sci. U.S.A.* **78**:143–146.

Kunkel, G. R., and Martinson, H. G., 1981, Nucleosomes will not form on double-stranded RNA or over poly(dA) · poly(dT) tracts in recombinant DNA, *Nucleic Acids Res.* **9**:6869–6888.

Kuo, M. T., Mandel, J. L., and Chambon, P., 1979, DNA methylation: Correlation with DNase I sensitivity of chicken ovalbumin and conalbumin chromatin, *Nucleic Acids Res.* **7**:2105–2114.

Larsen, A., and Weintraub, H., 1982, An altered DNA conformation detected by S1 nuclease occurs at specific regions in active chick globin chromatin, *Cell* **29**:609–622.

Lawson, G. M., Knoll, B. J., March, C. J., Woo, S. L. C., Tsai, J.-J., and O'Malley, B. W., 1982, Definition of 5' and 3' structural boundaries of the chromatin domain containing the ovalbumin multigene family, *J. Biol. Chem.* **257**:1501–1557.

Lomonossoff, G. P., Butler, P. J. G., and Klug, A., 1981, Sequence-dependent variation in the conformation of DNA, *J. Mol. Biol.* **149**:745–760.

Lowenhaupt, K., Keene, M. A., Cartwright, I. L., and Elgin, S. C. R., 1983a, Chromatin structure of eukaryotic genes: DNase I hypersensitive sites, *Stadler Genet. Symp.* **14** (in press).

Lowenhaupt, K., Cartwright, I. L., Keene, M. A., Zimmerman, J. L., and Elgin, S. C. R., 1983b, Chromatin structure in pre- and postblastula embryos of *Drosophila*, *Dev. Biol.* **99**:194–201.

Marshall, L. E., Graham, D. R., Reich, K. A., and Sigman, D. S., 1981, Cleavage of deoxyribonucleic acid by the 1,10-phenanthroline–cuprous complex: Hydrogen peroxide requirement and primary and secondary structure specificity, *Biochemistry* **20**:244–250.

Mayfield, J. E., Serunian, L. A., Silver, L. M., and Elgin, S. C. R., 1978, A protein released by DNase I digestion of *Drosophila* nuclei is preferentially associated with puffs, *Cell* **14**:539–544.

McGhee, J. D., Wood, W. I., Dolan, M., Engel, J. D., and Felsenfeld, G., 1981, A 200 base pair region at the 5' end of the chicken adult β-globin gene is accessible to nuclease digestion, *Cell* **27**:45–55.

Muskavitch, M., and Hogness, D. S., 1980, Molecular analysis of a gene in a developmentally regulated puff of *Drosophila melanogaster*, *Proc. Natl. Acad. Sci. U.S.A.* **77**:7362–7366.

Muskavitch, M. A. T., and Hogness, D. S., 1982, An expandable gene that encodes a *Drosophila* glue protein is not expressed in variants lacking remote upstream sequences, *Cell* **29**:1041–1051.

Nasmyth, K. A., 1982, The regulation of yeast mating-type chromatin structure by SIR: An action at a distance affecting both transcription and transposition, *Cell* **30**:567–578.

Nedospasov, S. A., and Georgiev, G. P., 1980, Non-random cleavage of SV40 DNA in the compact minichromosome and free in solution by micrococcal nuclease, *Biochem. Biophys. Res. Commun.* **92**:532–539.

Nickol, J., Behe, M., and Felsenfeld, G., 1982, Effect of the B–Z transition in poly(dG · m⁵dC) · poly(dG · m⁵dC) on nucleosome formation, *Proc. Natl. Acad. Sci. U.S.A.* **79**:1771–1775.

Palen, T., Gottschling, D. S., and Cech, T., 1982, Transcribed and non-transcribed regions of the ribosomal RNA gene of *Tetrahymena* exhibit different chromatin structures, *J. Cell Biochem. Suppl.* **6:**336.

Palmiter, R. D., Mulvihill, E. R., McKnight, G. S., and Senear, A. W., 1978, Regulation of gene expression in the chick oviduct by steroid hormones, *Cold Spring Harbor Symp. Quant. Biol.* **42:**639–648.

Peck, L. J., and Wang, J. C., 1981, Sequence dependence of the helical repeat of DNA in solution, *Nature (London)* **292:**375–378.

Que, B. G., Downey, K. M., and So, A. G., 1980, Degradation of deoxyribonucleic acid by a 1,10-phenanthroline–copper complex: The role of hydroxyl radicals, *Biochemistry* **19:**5987–5991.

Rhodes, D., and Klug, A., 1981, Sequence-dependent helical periodicity of DNA, *Nature (London)* **292:**378–380.

Samal, B., Worcel, A., Louis, C., and Schedl, P., 1981, Chromatin structure of the histone genes of *D. melanogaster, Cell* **23:**401–410.

Saragosti, S., Moyne, G., and Yaniv, M., 1980, Absence of nucleosomes in a fraction of SV40 chromatin between the origin of replication and the region coding for the late leader RNA, *Cell* **20:**65–75.

Selleck, S. B., Elgin, S. C. R., and Cartwright, I. L., 1984, Supercoil-dependent features of DNA structure at *Drosophila* locus *67B1* (submitted).

Sheffery, M., Rifkind, R. A., and Marks, P. A., 1982, Murine erythroleukemia cell differentiation: DNase I hypersensitivity and DNA methylation near the globin genes, *Proc. Natl. Acad. Sci. U.S.A.* **79:**1180–1184.

Shermoen, A. W., and Beckendorf, S. K., 1982, A complex of interacting DNase I-hypersensitive sites near the *Drosophila* glue protein gene, Sgs 4, *Cell* **29:**601–607.

Silver, L. M., and Elgin, S. C. R., 1976, A method for determination of the *in situ* distribution of chromosomal proteins, *Proc. Natl. Acad. Sci. U.S.A.* **73:**423–427.

Silver, L. M., and Elgin, S. C. R., 1978, Immunological analysis of protein distributions in *Drosophila* polytene chromosomes, in: *The Cell Nucleus V: Chromatin,* Part B (H. Busch, ed.), Academic Press, New York, pp. 216–263.

Sirotkin, K., and Davidson, N., 1982, Developmentally regulated transcription from *Drosophila melanogaster* chromosomal site 67B, *Dev. Biol.* **89:**196–210.

Sledziewski, A., and Young, E. T., 1982, Chromatin conformational changes accompany transcriptional activation of a glucose-repressed gene in *Saccharomyces cerevisae, Proc. Natl. Acad. Sci.* **79:**253–256.

Southern, E. M., 1975, Detection of specific sequences among DNA fragments separated by gel electrophoresis, *J. Mol. Biol.* **98:**503–517.

Stalder, J., Larsen, A., Engel, J. D., Dolan, M., Groudine, M., and Weintraub, H., 1980, Tissue-specific DNA cleavages in the globin chromatin domain introduced by DNase I, *Cell* **20:**451–460.

VanDyke, M. W., Hertzberg, R. P., and Dervan, P. B., 1982, Map of distamycin, netropsin, and actinomycin binding sites on heterogeneous DNA: DNA cleavage-inhibition patterns with methidiumpropyl-EDTA · Fe(II), *Proc. Natl. Acad. Sci. U.S.A.* **79:**5470–5474.

Weintraub, H., and Groudine, M., 1976, Chromosomal subunits in active genes have an altered conformation, *Science* **193:**848–856.

Weintraub, H., Larsen, A., and Groudine, M., 1981, α-Globin-gene switching during the development of chicken embryos: Expression and chromosome structure, *Cell* **24:**333–344.

Weintraub, H., Berg, H., Groudine, M., and Graf, T., 1982, Temperature-sensitive changes in the structure of globin chromatin in lines of red cell precursors transformed by a ts-AEV virus, *Cell,* **28:**931–940.

Wong, Y.-C., O'Connell, P., Rosbash, M., and Elgin, S. C. R., 1981, DNase I hypersensitive sites of the chromatin for *Drosophila melanogaster* ribosomal protein 49 gene, *Nucleic Acids Res.* **9:**6749–6762.

Wu, C., 1980, The 5' ends of *Drosophila* heat shock genes in chromatin are hypersensitive to DNase I, *Nature (London)* **286:**854–860.

Wu, C., and Gilbert, W., 1981, Tissue-specific exposure of chromatin structure at the 5' terminus of the rat preproinsulin II gene, *Proc. Natl. Acad. Sci. U.S.A.* **78:**1577–1580.

Wu, C., Bingham, P. M., Livak, K. J., Holmgren, R., and Elgin, S. C. R., 1979a, The chromatin structure of specific genes. I. Evidence for higher order domains of defined DNA sequence, *Cell* **16:**797–806.

Wu, C., Wong, Y.-C., and Elgin, S. C. R., 1979b, The chromatin structure of specific genes. II. Disruption of chromatin structure during gene activity, *Cell* **16:**807–814.

2

Expression of Transferred Thymidine Kinase Genes in Mouse L Cells: Evidence for Control by DNA Methylation

BARBARA CHRISTY and GEORGE SCANGOS

1. INTRODUCTION

The presence of methylated bases in the DNA of both prokaryotes and eukaryotes has been known for some time. In prokaryotic DNA, both 6-methyladenosine and 5-methylcytosine (m^5C) are found, while in mammalian DNA, the only detectable methylated base is m^5C (Wyatt, 1951). From 2 to 7% of cytosine residues are found as m^5C, and more than 90% of the methylated C residues are found in the dinucleotide sequence CpG (Grippo et al., 1970). The distribution of m^5C in the DNA is nonrandom, and both the pattern and level of methylation appear to be species-specific and tissue-specific (Razin and Riggs, 1980; Ehrlich and Wang, 1981). The pattern of methylation is probably established during development and is passed on from a differentiated cell to its daughters following the action of one or more "maintenance methylase" enzymes, which act on hemimethylated DNA shortly after replication (Gruenbaum et al., 1982). The symmetry of the CpG sequence allows

BARBARA CHRISTY and GEORGE SCANGOS • Department of Biology, The Johns Hopkins University, Baltimore, Maryland 21218.

one parental strand, which is methylated, to act as a template for the methylation of the new strand (Holliday and Pugh, 1975; Razin and Riggs, 1980; Ehrlich and Wang, 1981).

It has been proposed that methylation affects the processes of replication, transcription, and repair (J. H. Taylor, 1979). The best-characterized role of DNA methylation is in the control of gene expression through the control of transcription. A strong inverse correlation has been established between the amount of DNA methylation and the level of gene expression. When specific genes were examined [e.g., globin (van der Ploeg and Flavell, 1980), ovalbumin (Mandel and Chambon, 1979), ribosomal DNA (Bird *et al.,* 1981), immunoglobulin J chains (Yagi and Koshland, 1981) and metallothionein (Compere and Palmiter, 1981)], hypomethylation of the genes was seen in tissues in which the genes were expressed, while the same genes or their surrounding DNA, or both, were more heavily methylated in tissues in which they were not expressed. In several viral systems, endogenous unexpressed viral DNA was heavily methylated, while expressed viral DNA was relatively hypomethylated (Sutter and Doerfler, 1980; Vardimon *et al.,* 1980; Groudine *et al.,* 1981; Stuhlmann *et al.,* 1981; Harbers *et al.,* 1981). In a study of adenovirus gene expression, portions of an integrated genome were found to be methylated, while other portions of the genome were unmethylated. The level of methylation of the DNA of specific genes was inversely correlated with the level of expression of those genes (Sutter and Doerfler, 1980).

Data from various studies using 5-azacytidine (5-AC) to inhibit DNA methylation supports a causative role for DNA methylation in the inhibition of gene expression. 5-AC is an analogue of cytidine that becomes incorporated into DNA during replication, but that cannot be methylated (Jones and Taylor, 1980). The analogue contains a nitrogen atom in place of a carbon atom at the 5 position of the pyrimidine ring, which prevents it from accepting a methyl group at this position. The inability to accept methyl groups is not the only mechanism by which 5-AC inhibits methylation, however, since the degree of demethylation induced is greater than that which can be explained solely by the level of incorporation. The continued presence of 5-AC is not necessary for maintenance of the demethylated state (Jones and Taylor, 1980). Once demethylation has occurred, the pattern is maintained through successive generations. Groudine *et al.* (1981) demonstrated that endogenous viral genomes became hypomethylated and transcriptionally activated following growth of cells in the presence of 5-AC, Compere and Palmiter (1981) demonstrated that growth of cultured cells in the presence of 5-AC resulted in hypomethylation and activation of the metallothionein-I gene, and Clough *et al.* (1982) found that mouse cells transformed with herpes simplex virus (HSV) containing an inactive thymidine kinase *(TK)* gene could be induced to express the *TK* gene following growth in 5-AC. These and other results suggested a causative role for DNA methylation–demethylation in control of gene expression, but could not conclusively demonstrate that reexpression of the genes resulted directly from demethylation of gene sequences induced by 5-AC and not from undefined secondary effects.

Other types of experiments provided further support for the hypothesis that DNA methylation is causally related to the reduction of gene expression. Pollack *et al.* (1980) and Wigler *et al.* (1981) demonstrated that the HSV *TK* gene was reduced in its ability to transfer TK activity to Ltk⁻ cells when it had been partially methylated *in vitro* prior to transfer. TK⁺ cells that did arise contained copies of *TK* genes that were partially demethylated. These data suggested that methylation inhibited expression of the genes, but could not exclude the possibilities that the DNA methylation interfered with the transfer process itself, or that methylated DNA preferentially integrated into methylated inactive regions of the genome, or that both circumstances obtained.

Two recent studies provided partial answers to these objections. Waechter and Baserga (1982) reported that *in vitro* methylation of the *TK* gene at a site 79 base pairs (bp) upstream from the cap site reduced expression of the gene when microinjected into hamster cells. Although the authors used *Eco*RI methylase, which results in 5-methyladenine rather than m^5C normally found in mammalian cells, these data provided evidence that DNA methylation is sufficient to eliminate or abolish gene expression in hamster cells.

Stein *et al.* (1982) introduced genes for adenine phosphoribosyl transferase (APRT) into TK-deficit mouse L (Ltk⁻) cells by cotransformation with HSV TK and selection for *TK* expression. They found that *APRT* genes that had been methylated *in vitro* at CCGG sequences with *Hpa*II methylase were not expressed in recipient cells, while genes that were not methylated prior to introduction into cells were expressed. Furthermore, when cells that contained methylated, nonexpressed *APRT* genes were switched to medium in which *APRT* expression was required for survival, the rare APRT⁺ surviving cells were found to have demethylated *APRT* genes. These data again support the contention that DNA methylation is sufficient to abolish gene expression and that demethylation is a requisite for expression.

How does DNA methylation affect gene expression? One possibility is that the methyl groups interfere with DNA–protein interaction, since the methyl groups of 6-methyladenosine and m^5C protrude into the major groove of the DNA helix (Riggs, 1975). It has also been demonstrated *in vitro* that methylated C–G copolymers undergo a transition from the B to the Z state more readily than nonmethylated copolymers (Behe and Felsenfeld, 1981), so that methylated DNA may exist in an intracellular state different from unmethylated DNA under some conditions. Consistent with this interpretation, Kuo *et al.* (1979) found correlates between undermethylation, deoxyribonuclease (DNase) sensitivity, and expression of the ovalbumin gene. It is premature, however, to deduce any cause and effect between DNA methylation and chromatin structure as determined by DNase sensitivity.

2. RESULTS AND DISCUSSION

The HSV TK system is an excellent one for the examination of factors that regulate gene expression. The *TK* gene has been isolated on a 3.5-kilobase (kb) fragment of HSV DNA and has been cloned in pBR322 (Enquist *et al.,* 1979). A functional *TK* gene is not required for growth in normal tissue-culture medium, but agents selective for *TK* expression [hypoxanthine–aminopterin–thymidine (HAT)] (Littlefield, 1964; Szybalska and Szybalski, 1962) or for lack of expression [5-bromodeoxyuridine (BrdU)] can be added to the medium to create effective positive and negative selective systems. Finally, Ltk⁻ cells have an undetectable reversion rate (Scangos and Ruddle, 1981), so that they are excellent recipients. Thus, a purified gene can be transferred into Ltk⁻ cells; TK⁺ derivatives that have taken up and maintained and have expressed the gene can be recovered in HAT medium; TK-deficient derivatives of these can be selected in BrdU-containing medium; and TK⁺ reexpressing derivatives can again be selected by growth in HAT medium. The cloned *TK* gene can then be used as a probe to characterize changes in the state of the *TK* gene that correlate with changes in gene expression.

Some of the data presented below have been described in more detail (Christy and Scangos, 1982), and an independently derived cell line in which *TK* gene expression was regulated by similar mechanisms has recently been described (Ostrander *et al.,* 1982). In both these cell lines, *TK* gene expression appears to be regulated by methylation. When cell lines expressing *TK* were plated in medium containing BrdU to select TK⁻ derivatives, the *TK* genes in these derivatives were heavily methylated. Methylation, however, is not the universal response to selection against *TK* expression. In many cases, the genes were physically lost, and in others, the genes were retained with no detectable alterations in the state of methylation (our unpublished data; Robins *et al.*, 1981; Davies *et al.*, 1982). DNA methylation, therefore, is only one of several mechanisms by which gene expression can be regulated. The relative frequency with which the different mechanisms are used may be influenced by the copy number and site of integration of the transferred genes.

We describe below results obtained with one cell line, termed 101, and its derivatives. The 101 line is a TK$^+$ derivative of Ltk$^-$ that was obtained after DNA-mediated gene transfer of the circular plasmid pTKx-1 (Enquist *et al.*, 1979), containing the *TK* gene of HSV, into mouse Ltk$^-$ cells in the absence of carrier DNA. This cell line contains multiple copies of the plasmid arranged in a multimeric structure, probably at one chromosomal locus. The concatameric structure consists primarily of full-length plasmid molecules, but also includes deleted or rearranged copies of pTKx-1 (Huttner *et al.*, 1981; Christy and Scangos, 1982).

The 101 cells were plated in medium containing BrdU to select against expression of the *TK* gene, and TK$^-$ derivatives of 101 were isolated at a frequency of approximately 10^{-6}. One of these derivatives, 101BU1, was analyzed in detail. The 101BU1 cells contained no detectable *TK* activity and were unable to grow in selective (HAT) medium. Southern blot analysis revealed that 101BU1 cells retained the viral *TK* genes without any detectable rearrangements. Digestion of the DNAs of 101 and 101BU1 with *Hind*III, *Xba*I, *Bam*HI, and *Pvu*II failed to detect any difference in the pattern of TK$^-$ specific bands between the two lines (Christy and Scangos, 1982).

The 101BU1 cells were plated in HAT medium to select TK$^+$ derivatives that reexpressed the *TK* genes, and TK$^+$, HAT^r derivatives were obtained at a frequency of 5 $\times$ 10^{-7}. Two independent reexpressing cell lines (101H1 and 101HG) were isolated using this procedure and characterized further. A third TK$^+$ derivative (101HC) was isolated by treatment of 101BU1 with 5-AC prior to growth in HAT medium.

DNA was isolated from 101, 101BU1, 1011H1, 101HC, and 101HG and digested with the restriction enzymes *Msp*I and *Hpa*II. *Msp*I and *Hpa*II are isoschizomers, both of which recognize and cleave the sequence CCGG (Roberts, 1980). *Hpa*II cleaves only if the internal cytosine is unmethylated, while *Msp*I is insensitive to methylation at that site, so that a comparison of the pattern of bands visualized on Southern blots after digestion with the two enzymes provides an indication of the extent of DNA methylation. The plasmid pTKx-1 contains more than 40 sites for *Msp*I and *Hpa*II, yielding fragments of 9–600 bp (Sutcliffe, 1978; McKnight, 1981; Wagner *et al.*, 1981). Digestion of DNA from 101 or 101BU1 with *Msp*I generated a number of poorly resolved low-molecular-weight bands homologous to plasmid pTKx-1. Digestion of 101 DNA with *Hpa*II generated the same pattern of low-molecular-weight bands as well as a number of higher-molecular-weight bands, indicating that the *TK* genes in 101 were differentially methylated; some gene copies contained little or no methylation, while others were more heavily methylated. Digestion of 101BU1 with *Hpa*II yielded only high-molecular-weight bands—the low-molecular-weight bands seen in the *Msp*I digest of 101BU1 were not visualized. These data indicated that sites within the plasmid sequences that were not methylated in 101 had become methylated in the DNA of the TK-deficient cell line, 101BU1. The low-molecular-weight bands again were visualized in the *Hpa*II digests of DNA isolated from the reexpressing cell lines 101H1, 101HC, and 101HG, indicating that at least one copy of the *TK* gene in the DNA of these lines had become demethylated.

The pattern of high-molecular-weight bands seen after *Hpa*II digestion was similar in 101 and 101BU1; the major difference between the lines was the absence of low-molecular-weight bands in 101BU1 DNA. The presence of discrete high-molecular-weight bands indicated that at least some copies of the *TK* gene were somewhat methylated in 101 cells and that some *Hpa*II sites in the *TK* genes of 101BU1 cells were unmethylated. The fact that the pattern of bands was similar in the two lines indicated that the overall pattern of methylation had not been altered and that the hypermethylation in 101BU1 involved only a portion of the *TK* genes. This interpretation is supported by double digestion in which DNA from 101, 101BU1, and the three reexpressing cell lines was digested with both *Hind*III and *Hpa*II. All the *Hind*III bands were altered in mobility after the double digestions, indicating that none of the *Hind*III bands was composed of sequences that were methylated at every *Hpa*II site. The two lines have been maintained separately in culture for over 12 months, indicating that the pattern of methylation of the TK DNA is stable over time.

Because of the large number of *Hpa*II and *Msp*I sites located on the plasmid pTKx-1, no conclusions about specific sites of methylation within the *TK* gene could be drawn. For this reason, we used two other methylation-sensitive enzymes, *Sma*I and *Ava*I. *Sma*I recognizes five sites within pTKx-1 to generate bands of 0.2, 0.25, 0.8, 1.6, and 5.0 kb (M. Wagner and W.

Summers, personal communication). The 0.8- and 1.6-kb bands are derived entirely from HSV DNA; the sites that generate them lie in the 5' flanking region, within the structural gene, and in the 3' flanking DNA. The 5.0-kb band is made up primarily of pBR322 and a small amount of TK DNA. The 0.8-, 1.6-, and 5.0-kb bands could be seen clearly in Southern blots of 101 DNA after *Sma*I digestion. All three bands were undetectable in 101BU1 DNA and reappeared in the DNA of 101H1, 101HC, and 101HG. *Ava*I, which is also sensitive to CpG methylation, recognizes eight sites in pTKx-1 (M. Wagner and W. Summers, personal communication). After *Ava*I diges-tion, two bands of 0.64 and 0.78 kb. derived from within the *TK* gene, were present in the DNA of 101 and all the reexpressors, but absent in the DNA from 101BU1. These data demonstrated that at least one copy of the *TK* gene within the DNA of each of the four TK $^+$ lines contained unmethylated *Hpa*II, *Sma*I, and *Ava*I sites and that those same sites were methylated in the DNA of 101BU1.

The restriction-enzyme data demonstrated an inverse correlation between the level of methylation and expression of the transferred *TK* genes. If DNA methylation is causally related to the elimination of *TK* gene expression, then inhibition of DNA methylation in 101BU1 by growth in the presence of 5-AC should increase the frequency of TK $^+$ derivatives. Growth of 101 cells in 5-AC decreased the overall extent of DNA methylation in the cells as judged by ethidium bromide staining of *Msp*I- and *Hpa*II-digested DNAs following electrophoresis. Growth of 101BU1 (TK $^-$) cells in 2, 5, or 10 μM 5-AC for 2–5 days prior to HAT selection resulted in a 6- to 23-fold increase in the number of TK $^+$ colonies compared to untreated control cells. Treatment with 5 μM 5-AC for 3–4 days produced a maximal increase in *TK* reexpres-sion, while cell death due to toxicity of 5-AC was greatest in 10 μM 5-AC. The reexpressor cell line 101HC, derived from 101BU1 following treatment with 5-AC and HAT selection, contained *TK* genes that were amplified and hypomethylated relative to 101BU1. The fact that the *TK* genes in 101BU1 became reexpressed following treatment with 5-AC supports a causal rela-tionship between methylation and elimination of gene expression. However, as in previous studies in which 5-AC reactivation of various genes has been noted (Jones and Taylor, 1980; Compere and Palmiter, 1981; Groudine *et al.*, 1981), these data do not unambiguously determine whether demethylation of the gene and flanking sequences or demethylation of distal genes that somehow affected transcription was the important event in gene reexpression. It is even possible that mutational events induced by 5-AC could be the important factor, although it has recently been shown by Landolph and Jones (1982) that 5-AC was not significantly mutagenic in $10T_{1/2}$ cells.

For this reason, DNA samples isolated from 101 and 101BU1 and from each of the reexpressing cell lines were used in secondary gene transfer. If DNA methylation directly prevents expression of the *TK* genes in 101BU1, then DNA isolated from 101BU1 should not be able to transfer the TK$^+$ phenotype in secondary gene-transfer experiments. DNA was isolated from 101, 101BU1, and the three reexpressors and applied to mouse Ltk$^-$ cells by calcium phosphate precipitation. TK$^+$ colonies were obtained when DNA from 101 or any of the reexpressors was used. No TK$^+$ colonies were obtained when DNA from 101BU1 was used as the donor under conditions where 101 DNA generated 355 TK$^+$ colonies. Several independent preparations of 101BU1 DNA were used to eliminate the possibility that the lack of ability to transfer the TK$^+$ phenotype was merely a function of a "poor" DNA preparation. Since gene transfer is thought to separate the TK DNA from the rest of the genome and other cellular constituents, the difference in *TK* expression in lines 101 and in 101BU1 probably lies within the *TK* genes or flanking DNA. The only detectable difference between the *TK* genes in lines 101 and 101BU1 was the extent of DNA methylation. These data support the hypothesis that the elimination of *TK* gene expression in 101BU1 was due to hypermethylation of the *TK* genes or closely flanking DNA.

Although the structure of the TK-specific DNA in 101BU1 is not detectably different from that in 101, each of three TK$^+$ reexpressors examined had alterations in the pattern or intensity of TK-specific bands after digestion with *Pvu*II, *Bam*HI, or *Hind*III. DNA from 101H1 contained an additional band of about 10 kb after *Xba*I digestion and two additional bands after *Hind*III digestion. In addition, *Hind*III bands of 17 and 5 kb were reduced in intensity relative to other bands. DNA from 101HG was missing all but the 7.9- and 3.5-kb *Hind*III bands. These data indicated that both 101H1 and 101HG had suffered deletions of the inserted TK DNA. DNA of 101HC had band patterns indistinguishable from that of 101 and 101BU1 following *Hind*III digestion, except that the intensity of the bands was increased, suggesting that the region of TK DNA in this line had undergone an amplification.

To determine whether the alterations in gene structure were correlated with the changes in *TK* gene expression, 12 subclones of 101BU1 were isolated randomly without HAT selection. Following *Hind*III digestion, DNA from 7 subclones was indistinguishable from 101BU1 DNA, while DNA from 5 subclones exhibited changes in the pattern of bands homologous to pTKx-1.

The existence of rearrangements in 5 of 12 random subclones indicates that rearrangement in the region of the introduced *TK* genes is a common event.

We have not yet determined whether the gene rearrangements are necessary for demethylation or reexpression. It may be that some rearrangements allow reexpression by bringing the methylated *TK* genes into proximity with hypomethylated DNA, thereby disrupting the pattern of methylation of the *TK* genes. Alternatively, demethylation may occur without rearrangement, but not in itself be sufficient for reexpression. Subsequent rearrangements may be necessary for expression. Finally, DNA rearrangements may occur randomly, without relation to gene expression. Experiments to determine the nature of the rearrangements in the reexpressors are in progress.

ACKNOWLEDGMENT. This work was supported by NSF award PCM8105296. This is contribution number 1211 from the Department of Biology, The Johns Hopkins University.

REFERENCES

Behe, M., and Felsenfeld, G., 1981, *Proc. Natl. Acad. Sci. U.S.A.* **78**:1619–1623.

Bird, A., Taggert, M., and Macleod, D., 1981, *Cell* **26**:381–390.

Breznik, T., and Cohen, J. C., 1982, *Nature (London)* **295**:255–257.

Christy, B., and Scangos, G., 1982, *Proc. Natl. Acad. Sci. U.S.A.* **79**:6299–6303.

Clough, D. W., Kunkel, L. M., and Davidson, R. L., 1982, *Science* **206**:70–73.

Compere, S. J., and Palmiter, R. D., 1981, *Cell* **25**:233–240.

Davies, R. L., Fuhrer-Krusi, S., and Kucherlapati, R. S., 1982, *Cell* **31**:521–529.

Ehrlich, M., and Wang, R. Y.-H., 1981, *Science* **212**:1350–1357.

Enquist, L. W., VandeWoude, G. F., Wagner, M., Smiley, J. R., and Summers, W. C., 1979, *Gene* **7**:335–342.

Grippo, P., Iacarino, M., Parisi, E., and Scarano, E., 1970, *J. Mol. Biol.* **35**:195–208.

Groudine, M., Eisenman, R., and Weintraub, H., 1981, *Nature (London)* **292**:311–317.

Gruenbaum, Y., Cedar, H., and Razin, A., 1982, *Nature (London)* **295**:620–622.

Harbers, K., Schnieke, A., Stuhlmann, H., Jahner, D., and Jaenisch, R., 1981, *Proc. Natl. Acad. Sci. U.S.A.* **78**:7609–7613.

Holliday, R., and Pugh, J. E., 1975, *Science* **187**:226–232.

Huttner, K. M., Barbosa, J. A., Scangos, G. A., and Ruddle, F. H., 1981, *J. Cell Biol.* **91**:153–156.

Jones, P. A., and Taylor, S. M., 1980, *Cell* **20**:85–93.

Kuo, M. T., Mandel, J. L., and Chambon, P., 1979, *Nucleic Acids Res.* **7**:2105–2113.

Landolph, J. R., and Jones, P. A., 1982, *Cancer Res.* **42**:817–823.

Littlefield, J. W., 1964, *Science* **145**:709–710.

Mandel, J. L., and Chambon, P., 1979, *Nucleic Acids Res.* **7**:2081–2103.

McKnight, S. L., 1981, *Nucleic Acids Res.* **8**:5949–5964.

Ostrander, M., Vogel, S., and Silverstein, S., 1982, *Mol. Cell. Biol.* **2**:708–714.

Pollack, Y., Stein, R., Razin, A., and Cedar, H., 1980, *Proc. Natl. Acad. Sci. U.S.A.* **77**:6463–6467.
Razin, A., and Riggs, A. D., 1980, *Science* **210**:605–610.
Riggs, A., 1975, *Cytogenet. Cell Genet.* **14**:9–25.
Roberts, R., 1980, *Gene* **8**:329.
Robins, D. M., Axel, R., and Henderson, A. S., 1981, *J. Mol. Appl. Genet.* **1**:191–203.
Scangos, G. A., and Ruddle, F. H., 1981, *Gene* **14**:1–10.
Stein, R., Greenbaum, Y., Pollack, Y., Razin, A., and Cedar, H., 1982, *Proc. Natl. Acad. Sci. U.S.A.* **79**:61–65.
Stuhlmann, H., Jahner, D., and Jaenisch, R., 1981, *Cell* **26**:221–232.
Sutcliffe, J. G., 1978, *Nucleic Acids Res.* **5**:2721–2728.
Sutter, D., and Doerfler, W., 1980, *Proc. Natl. Acad. Sci. U.S.A.* **77**:253–256.
Szybalska, E. H., and Szybalski, W., 1962, *Proc. Natl. Acad. Sci. U.S.A.* **48**:2026–2034.
Taylor, J. H., 1979, in: *Molecular Genetics,* Pt. 3, *Chromosome Structure* (J. H. Taylor, ed.), Academic Press, New York, pp. 89–115.
Van der Ploeg, L. H. T., and Flavell, R. A., 1980, *Cell* **18**:947–958.
Vardimon, L., Neuman, R., Kuhlmann, I., Sutter, D., and Doerfler, W., 1980, *Nucleic Acids Res.* **8**:2461–2473.
Waechter, D. E., and Baserga, R., 1982, *Proc. Natl. Acad. Sci. USA* **79**:1106–1110.
Wagner, M. J., Sharp, J. A., and Summers, W. C., 1981, *Proc. Natl. Acad. Sci. U.S.A.* **78**:1441–1445.
Wigler, M., Levy, D., and Perucho, M., 1981, *Cell* **24**:33–40.
Wyatt, G. R., 1951, *Biochem. J.* **48**:584–590.
Yagi, M., and Koshland, M. E., 1981, *Proc. Natl. Acad. Sci. U.S.A.* **78**:4907–4911.

3

Pattern of Histone-Variant Synthesis and Implications for Gene Regulation

ROY S. WU and WILLIAM M. BONNER

1. INTRODUCTION

Studies on histone proteins impinge in two different ways on the study of transcription: (1) the role of histone proteins in the structure and function of chromatin and (2) the regulation of histone gene expression. It is generally accepted that histones serve a structural role in the packaging of the DNA in the cell nucleus (Felsenfeld, 1978). Thus, histones may affect gene expression by changing the structure of chromatin. However, the many complex and multiple posttranslational modifications of histones suggest that they have other roles in addition to purely structural ones.

It was generally believed that the synthesis of histone proteins was tightly coupled to DNA replication and thereby limited to the S phase of cycling cells (for a review, see Elgin and Weintraub, 1975). This rather simple model has been complicated by two recent findings. First, some of the five histone

ROY S. WU • Biotech Research Laboratories, Inc., Rockville, Maryland 20850; Laboratory of Molecular Pharmacology, Division of Cancer Treatment, National Cancer Institute, National Institutes of Health, Bethesda, Maryland 20205.   WILLIAM M. BONNER • Laboratory of Molecular Pharmacology, Division of Cancer Treatment, National Cancer Institute, National Institutes of Health, Bethesda, Maryland 20205.

proteins were found to be composed of slightly different sequences, known as isoproteins or variants (Marzluff *et al.*, 1972; Franklin and Zweidler, 1977; Von Holt *et al.*, 1979; West and Bonner, 1980a). These results indicate that histones are members of multigene families. Second, specific subsets of histone variants were shown to be synthesized during different periods of the cell cycle (R. S. Wu and Bonner, 1981) and under different physiological states (R. S. Wu *et al.*, 1982b, 1983a). Some of this histone synthesis is not linked to DNA synthesis. Thus, the regulation of histone gene expression is far more complex than originally envisioned.

In this chapter, we will attempt to relate the implications of these new findings on the existence and synthesis of histone variants for both the areas of chromatin structure and gene expression and the control of histone gene expression.

2. HISTONE PROTEINS: CLASSIFICATION OF VARIANTS

The histone proteins fall into five major classes, H1, H2A, H2B, H3, and H4; the latter four histones are known as core histones because they are found in the core particle of nucleosomes. All five histones are small basic proteins that are acid-extractable and lack tryptophan. Although the amino acid sequences of the histones as a general class of proteins have been highly conserved during evolution, variations in histone sequences do occur (Isenberg, 1979; Von Holt *et al.*, 1979). Sequence variations fall into two classifications. First, sequences have varied during the evolution of eukaryotic species. However, despite these changes in the amino acid composition, the standard free energy of binding (Klotz *et al.*, 1975) between histones derived from the same species or different species remains relatively constant (D'Anna and Isenberg, 1974; Spiker and Isenberg, 1977a,b; Mardian and Isenberg, 1978). Such lack of deviation suggests that the binding surfaces between histones have been highly conserved. Second, sequences have varied within the same species; these can be further divided into three groups. One group of variants is associated with changes occurring over time, such as during development; thus, there are early and late histone variants in certain organisms (Alfageme *et al.*, 1974; Newrock *et al.*, 1978; R. S. Wu *et al.*, 1982a). A second group belongs to changes arising from terminal differentiation of specific tissues; thus, there are sperm-specific (Von Holt *et al.*, 1979) and red-blood-cell specific variants (Hnilica, 1964; Neelin *et al.*, 1964). Finally, there is a group of variants found in the same nucleus at the same time (West and Bonner, 1980a).

All the histone variants can undergo posttranslational modifications (for a review, see Isenberg, 1979) such as acetylation, methylation, phosphorylation, ubiquitination, and poly ADP-ribosylation. Some modifications, such as phosphorylation of H1 (Bradbury *et al.*, 1973; Gurley *et al.*, 1978) and ubiquitination of H2A and H2B (Matsui *et al.*, 1979; R. S. Wu *et al.*, 1981), correlate with cell-cycle events, but the function of most modifications is still obscure.

2.1. Histone-Variant Separation and Characterization

Many histone variants and their modified forms may be resolved and separated from each other by taking advantage of the interaction of these proteins with nonionic detergents as described by Alfageme *et al.* (1974) and Franklin and Zweidler (1977). Additional separation from nonhistone proteins and improved resolution can be obtained using the two-dimensional (2D) gel system described by Bonner *et al.* (1980). A typical 2D gel of an acid extract obtained from salt-washed nuclei of L1210 cells is shown in Fig. 1. Salt washing removes most of the nonhistone proteins; in extracts of whole nuclei (Bonner *et al.*, 1980), many nonhistones remain, forming a diagonal. With 8 M urea in the first dimension, these nonhistones may interfere with H2B and H4. By varying the urea concentration in the resolving gel, the interaction between the core histones and the nonionic detergent can be modulated, resulting in increased or decreased resolution of specific groups of histone variants (West and Bonner, 1980b) as well as improved separation of histones from nonhistones. For example, with 6 M urea (Fig. 2) instead of 8 M urea (Fig. 1) in the first dimension, the H4 and the H2Bs are resolved from the diagonal of nonhistone proteins, whereas the resolution of the H2A.1/.2 region is decreased. In this regard, it should be mentioned that there could also be variants yet to be discovered, since the resolution depends on Triton binding. If two variants bind Triton similarly, they may not be resolved. However, our evidence from peptide-mapping studies (West and Bonner, 1980a,b; Pantazis and Bonner, 1981; R. S. Wu *et al.*, 1981, 1982a) indicates that each variant spot contains a homogeneous protein sequence.

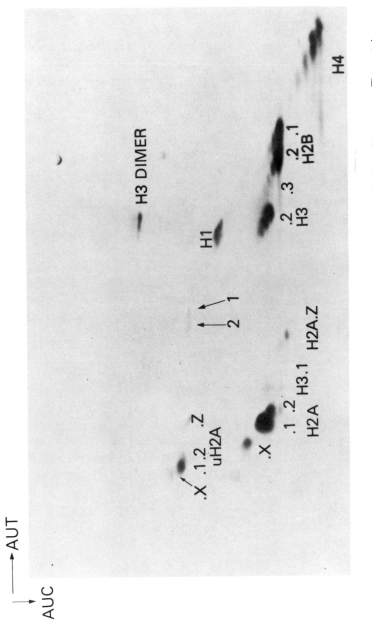

FIGURE 1. Two-dimensional analysis of the acid-soluble proteins from mouse L1210 nuclei after salt treatment. The proteins were electrophoresed in a first-dimension (1D) acetic acid–urea–Triton X-100 (AUT) gel with 8 M urea and 8 mM Triton, then in a second-dimension (2D) acetic acid–urea–cetyltrimethylammonium bromide (AUC) gel. The proteins were stained with Coomassie blue. The arrowed numbers 1 and 2 denote uH2Bs. From West and Bonner (1980b).

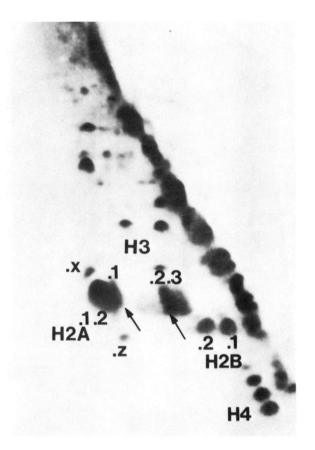

FIGURE 2.   Two-dimensional analysis of the acid-soluble proteins from mouse 3T3 cells. The [^{14}C]arginine radioactive proteins were electrophoresed in a 1D AUT gel with 6 M urea and 8 mM Triton, then in a 2D AUC gel. (←) Locations of H3.1 and H3.2 variants. Gels were fluorographed for 24 hr.

2.1.1. Histone 4

H4 has one variant with four possible modified forms. H4, an arginine-rich histone, is one of the most conserved proteins known. For example, H4 of pea plants differs from H4 of calf thymus by only two conservative substitutions. The H4 from sea urchin embryos contains a Cys residue instead of Ser, as in mammalian H4s, and migrates more slowly in acid–urea–Triton X-100 gels (Pantazis and Bonner, 1981; R. S. Wu *et al.*, 1982a).

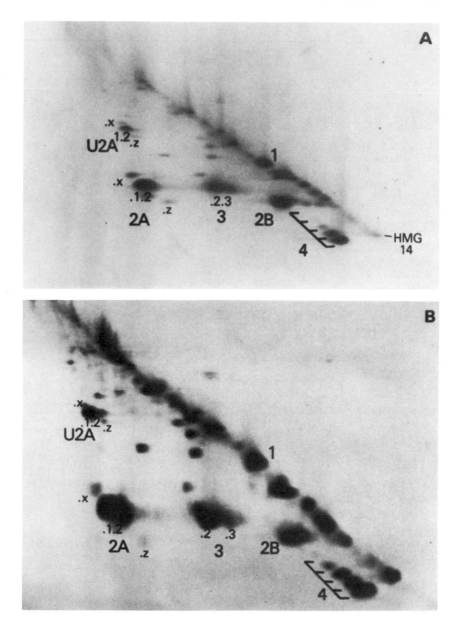

FIGURE 3.   Accumulation of radioactivity in histones of CHO cells. (A) Mass pattern of acid extract obtained from whole cells in G_1; S-phase cells yield the same pattern; (B) fluorogram of S-phase cells; (C) fluorogram of G_1-phase cells. From R. S. Wu and Bonner (1981).

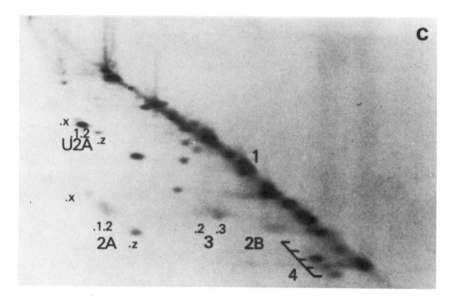

FIGURE 3. *Continued*

2.1.2. Histone 3

H3, another arginine-rich histone, is also highly conserved. Four amino acid differences have been detected between pea and calf. Despite the limited number of amino acid differences recorded among species, the H3s from different organisms migrate differently on acid–urea–Triton X-100 gels (R. S. Wu and W. M. Bonner, unpublished data). As shown in Figs. 1 and 2, mammalian H3 resolves into three variants, 3.1, 3.2, and 3.3. The H3 variants differ by one or two amino acids (Urban *et al.*, 1979); H3.1 has Cys at position 96, while H3.2 and H3.3 have Ser (Marzluff *et al.*, 1972). H3.1 and H3.2 have Val-Met at positions 89-90; H3.3 has Ile-Gly (Franklin and Zweidler, 1977).

The amount of each variant may vary from cell to cell and tissue to tissue (Zweidler, 1976). For example, H3.3 is the major variant in mouse liver, while the H3.2 variant is the major variant in mouse L1210 cells. Humans and cows have much more H3.1 than H3.2. Zweidler (1980) reported that the mass fraction of H3 in H3.3 increases from 15% in livers of newborn mice to 55% in livers of adult mice, showing that variant patterns may change with age. Recently R. S. Wu *et al.* (1983a) reported that H3.3 is the major

H3 variant in quiescent human lymphocytes. However, by three days after activation with phytohemagglutinin, H3.1 and H3.2 have become the major H3 variants.

2.1.3. Histone 2B

H2B is evolutionarily more divergent than the H3s and H4. The carboxyl two thirds of the molecule is highly conserved, while the amino one third is more variable. H2B can be resolved into two variants in mouse cells, H2B.1 and H2B.2 (see Fig. 1). In other cell lines such as IMR-90, a human embryonic lung fibroblast line, and Chinese hamster ovary (CHO) cells, there is only one detectable H2B (Fig. 3).

2.1.4. Histone 2A

The number of H2A variants was not clearly established until recently (Urban *et al.,* 1979; Allis *et al.,* 1980; Palmer *et al.,* 1980; West and Bonner, 1980a). West and Bonner (1980a) showed in mammals that H2A has the most variant forms of all the core histones. The four H2A variants can be classified into two groups, two homeomorphous variants, H2A.1 and H2A.2, and two heteromorphous variants, H2A.X and H2A.Z. The heteromorphous variants always represent only 5–20% of the total H2As in mammalian cells. The four proteins were shown to be H2A variants by several criteria. All four proteins contained the conserved H2A peptide sequence Ala-Gly-Leu-Gln-Phe-Pro-Val-Gly-Arg. The Lys/Arg ratios of the four proteins are similar to each other, but very different from those of other core histones. All four proteins can be modified by the covalent attachment of ubiquitin (West and Bonner, 1980a).

H2A.1 and H2A.2 are very similar. There are changes at residue 16 of Thr (H2A.1) to Ser (H2A.2) (Franklin and Zweidler, 1977), at residue 51 of Leu (H2A.1) to Met (H2A.2) (Sautiere *et al.,* 1975), at residue 99 of Arg (H2A.1) to Lys (H2A.2) (Laine *et al.,* 1976), and at residue 41 of Glu (H2A.1) to Asn (H2A.2) (Isenberg, 1979; West and Bonner, 1983).

With sodium dodecyl sulfate (SDS)–polyacrylamide gel analysis of mouse H2As, the apparent molecular weight of H2A.X is 1000 daltons larger, and that of H2A.Z is 600 daltons smaller, than that of H2A.1 (West and Bonner, 1980a). Using H2A.1 as reference, H2A.X was found to have large regions of homology. The regions of nonidentity in H2A.X usually coincided with known regions of variability in sequenced H2As. H2A.X has one Arg and six Lys peptides not found in H2A.1. H2A.X has the N-terminal Ac-Ser-Gly-

Arg like H2A.1 and H2A.2 (Pantazis and Bonner, 1981), but does not contain sulfur-containing amino acids in any of its peptides (West and Bonner, 1980a).

H2A.Z, like its H2A.X counterpart, does not have Met or Cys in its sequence. However, it differs considerably from all the other H2As. It has only two Arg peptides and no Lys peptides in common with H2A.1 (Pantazis and Bonner, 1981; R. S. Wu *et al.*, 1982a). Furthermore H2A.Z does not have the acetylated Ser at the N terminus and does not seem to be phosphorylated (Pantazis and Bonner, 1981). Recently Ball *et al.* (1983) have sequenced the first 30 amino acids of H2A.Z from calf thymus. Only 60% sequence homology relative to H2A.1 was retained in the N-terminal domain. H2A.Z has a two-residue extension and 10 amino acid substitutions all occurring within the first 21 amino acid residues. These substitutions result in a change in the net charge of the N-terminal domain in H2A.Z to $+6$ as opposed to $+8$ for the corresponding region of H2A.1.

When the H2A proteins from two different organisms, mouse and sea urchin, were mixed and analyzed on the 2D gel system of Bonner *et al.* (1980), H2A.Z from mouse comigrated with H2A.Z from sea urchin, whereas all the other H2As migrated differently (R. S. Wu *et al.*, 1982a). Peptide mapping demonstrated that H2A.2 (mouse) and H2A.α (sea urchin) differed at many positions. There were five Arg but only two Lys peptides in common. One difference was the substitution of Phe (sea urchin) for a Leu (mouse) in one of the Arg peptides, giving sea urchin H2A.α two Phe residues instead of one. Most of the differences were due to known sequence differences across species (Isenberg, 1979). On the other hand, H2A.Z from mouse and H2A.Z from sea urchin had seven Arg and eight Lys peptides in common. Thus, the H2A.Z variant is clearly much more conserved evolutionarily than the other major H2As; yet at the same time, it is quite different from other H2As in both these organisms. In fact, the conservative nature of H2A.Z sequences is more like that of the H4s and the H3s. Another interesting feature of H2A.Z from sea urchin is that it is the only H2A variant that does not seem to be developmentally stage-specific; it is synthesized throughout early development (R. S. Wu *et al.*, 1982a).

2.1.5. Histone 1

H1 is the most variable of the histones. The number of variants changes from tissue to tissue, and for a given tissue, it varies from one species to another (Gurley *et al.*, 1975; Isenberg, 1979; Smith and Johns, 1980). One H1 variant, H1^0, is thought to be specific for stationary cells. H1 interacts little, if at all, with nonionic detergents and thus remains on the diagonal in the 2D gel of Bonner *et al.* (1980). H5, a histone that is found only in avian red cells, is a close relative of H1 (Hnilica, 1964; Neelin *et al.,* 1964).

2.2. Histone-Variant Synthesis Patterns

The ultimate result of most gene expression is the synthesis of protein. Whatever hypotheses are proposed for the regulation of histone gene expression must reflect what is seen on the protein level. This laboratory has studied the patterns of synthesis of the histone variants in cycling cells (R. S. Wu and Bonner, 1981) and in cells in different physiological states (R. S. Wu *et al.,* 1982b, 1983a). We review some of these findings and present some recent results on the synthesis pattern of histone variants in quiescent and transformed cells.

2.2.1. Cycling Cells

Figure 3, from R. S. Wu and Bonner (1981), shows the synthesis patterns of histone variants in CHO cells during the G_1 and S phases of the cell cycle. The histone synthesis pattern in G_1 differed quantitatively and qualitatively from that in S. The qualitatively different pattern precludes contamination by S-phase cells as an explanation for G_1 histone synthesis. Assays (Groppi and Coffino, 1980; Delegeane and Lee, 1981; Marashi *et al.,* 1982) that did not resolve variants could not preclude this explanation and therefore could not definitely prove histone synthesis in non-S phases of the cell cycle.

Certain histone variants that were synthesized as minor components in S accounted for much larger fractions in G_1 (Fig. 3). These variants were the H2A variants .X and .Z and the H3 variant .3. The small amount of H2A.1 and H2A.2 that is synthesized during G_1 has been attributed to contamination by a small number of S-phase cells (1.4% of total nuclei were labeled with [^{3}H]thymidine during G_1).

A detailed time–course analysis in which the radioactivity in various histones was quantitated is shown in Fig. 4. The synthesis of H2A.1 and H2A.2, as well as that of H2B and H4, increased proportionately with the increase in DNA synthesis. In contrast, the synthesis of H2A.X and .Z remained relatively constant. In terms of total H2A synthesis, H2A.X accounted for 38% before mitosis (-1 hr), but its share dropped to about 8% in late G_1 (4 hr). The synthesis of H2A.Z remained at 23% in G_2 (-1 hr) and G_1 (2 hr), but dropped to 3% in S phase (6–8 hr), while synthesis of H2A.1 and

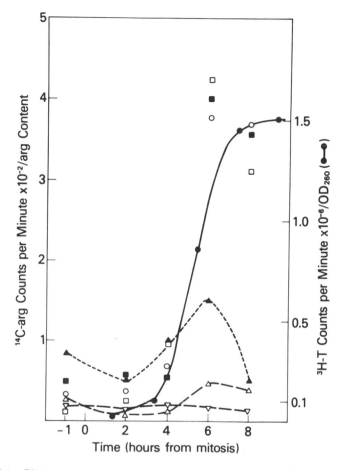

FIGURE 4. DNA and protein synthesis during the cell cycle. (●—●) DNA; (□—□) H2A.1/.2; (■—■) H2B; (○—○) H4; (▲---▲) HMG 14; (△---△) H2A.X; (▽---▽) H2A.Z. $G_2 = -1$ hr; mitosis = 0 hr, $G_1 = 0$–3.5 hr, S = 3.5–8 hr. From R. S. Wu and Bonner (1981).

H2A.2 increased to 80% during S phase. In other experiments, the H3-variant synthesis pattern was similarly analyzed. The H3.1- and H3.2-variant synthesis pattern correspond closely to the H2A.1 and H2A.2 synthesis pattern. In CHO cells, the H3.2 synthesis represented 68% and the H3.1 synthesis approximately 20% of the total H3 synthesis during S. The H3.3 synthesis accounted for more than 80% of the synthesis during G_1 and G_2 and dropped to 15% of the total H3 synthesis during S.

   The synthesis of the four core histones is maintained at close to nucleo-somal ratios throughout the cell cycle. This suggests that there are two types

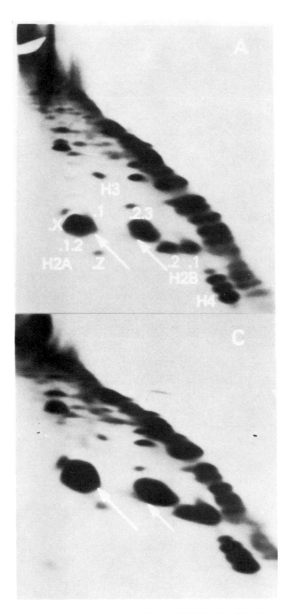

FIGURE 5.   Histone-variant synthesis in 3T3 and A31T cells. (A) Exponential 3T3; (B) quiescent 3T3; (C) exponential A31T; (D) overgrown A31T. Arrows denote the positions of the H3.1 and H3.2 variants.

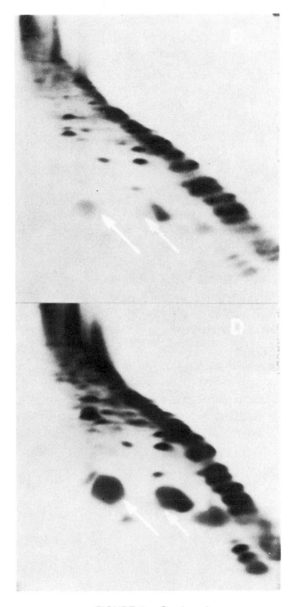

FIGURE 5. *Continued*

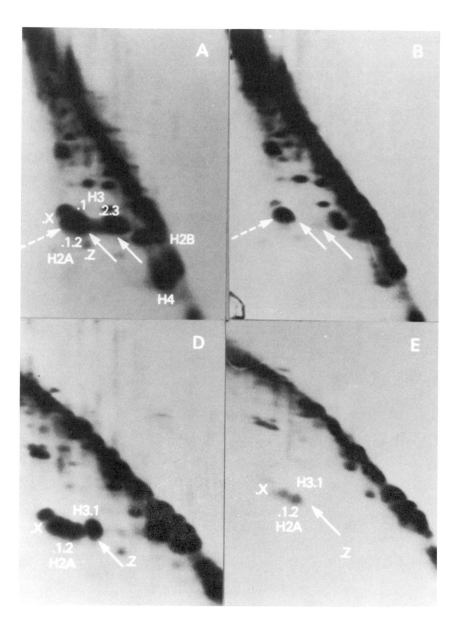

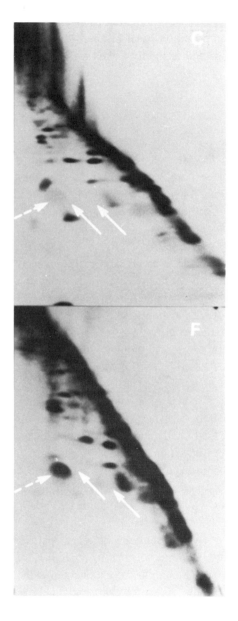

FIGURE 6. Histone-variant synthesis pattern of CHO cells in various growth states. (A, D) Exponential; (B, E) quiescent; (C) G_1; (F) quiescent treated with hydroxyurea. The first-dimensional gel in (A, B, C, F) was 15% acrylamide, 6 M urea, 8 mM Triton. The first-dimensional gel in (D, E) was 10% acrylamide, 8 M urea, 8 mM Triton, which increases the resolution of the H2A/H3.1 region. Gels in (A, D) were exposed for 1 week and gels in (B, C, E, F) for 3 weeks. (⟶) Positions of the H3.1 and H3.2 variants; (----→) positions of the H2A.1 and H2A.2 variants.

of histone synthesis in cycling cells. One type, named basal histone synthesis, occurs throughout the cycle and is not linked to DNA replication. Basal histone synthesis is not inhibited when DNA synthesis is inhibited by hydroxyurea (R. S. Wu and Bonner, 1981). Superimposed on basal histone synthesis is S-phase synthesis, which increases with and is tightly linked to replicative DNA synthesis.

2.2.2. Quiescent Cells

Another physiological state besides G_1 during which DNA synthesis is greatly reduced is the quiescent state, G_0. Mouse 3T3 cells are known to contact-inhibit (Abercrombie and Ambrose, 1962) rather strongly when they grow to confluency (Todaro *et al.*, 1965; for a review, see Holley, 1975). Figure 5A shows the pattern of histone synthesis during exponential growth. The H3.2 variant is the dominant species of H3 being synthesized; H3.3 and H3.1 are synthesized in lesser amounts. All the H2As, H2Bs, and H4 species are synthesized. After confluency is reached, H3.2- and H3.1-variant synthesis is not detectable (Fig. 5B). On refeeding a confluent culture, in which the H3.1 and .2 synthesis is undetectable, these variants are synthesized as in exponential cells (data not presented). A31T, a spontaneous transformant of 3T3, does not contact-inhibit but continues to grow in humps of cells. The synthesis of H3.2 and H3.1 continues in overgrown cultures, although their fraction of the total H3 is noticeably decreased (Fig. 5D) when compared to the exponential control (Fig. 5C).

To demonstrate that histone synthesis seen in quiescent cells is a general phenomenon and to relate quiescent pattern of synthesis to basal synthesis seen in G_1, exponential and quiescent CHO cells and resting lymphocytes were utilized. Figure 6 shows the pattern of synthesis of CHO cells during exponential growth and quiescence. During exponential growth, all variants are synthesized. The H3-variant synthesis pattern is the same for both quiescent and G_1 cells; H3.1 and H3.2 are not synthesized, while the H3.3 variant is synthesized.

However, the H2A-variant pattern distinguishes the quiescent pattern from the basal synthesis pattern found in G_1. In quiescent cells, the four H2A variants are synthesized in similar relative amounts as in exponential cells; indeed, the (H2A.1 + .2/H2A.X + .Z) ratio of 3.5 for exponential cells is not significantly different from that of G_0 cells (3.9). When the data in Figs. 4 and 6 are used to calculate the H2A-variant synthesis ratio (H2A.1 + .2/H2A.X + .Z) during the transition from G_1 to S, the ratio increases from approximately 0.4 at 1 hr after shakeoff (G_1) to 5.0 at 6 hr, well into S phase. Thus, from the standpoint of H2A-variant synthesis, the quiescent pattern of histone-variant synthesis is not the same as any pattern in G_1 or at the G_1–S boundary. These data, in conjunction with the H3-variant synthesis data discussed above, strongly suggest that the quiescent state is a separate and discrete state rather than a part of G_1.

In the case of physiologically normal resting cells, the T lymphocytes, a significant level of quiescent histone synthesis was detected (R. S. Wu *et al.*, 1983a). H3.3 was the major H3 variant synthesized as well as being the dominant mass variant. When T lymphocytes were induced to proliferate, the histone synthesis pattern quickly changed to the S-phase pattern; H3.1 and H3.2 synthesis dominated. Gradually over a three-day period the H3 mass pattern changed until H3.1 and H3.2 became the dominant species. These results extended the previous findings (R. S. Wu *et al.*, 1982b) in three ways. First, since T lymphocytes are a normal bodily constituent, quiescent histone synthesis is not an aberration of cells cultured *in vitro* but a characteristic of normal quiescent cells. Second, T lymphocytes are in suspension and not attached to a surface; quiescent histone synthesis is not the result solely of contact inhibition (Todaro *et al.*, 1965), but more generally of a quiescent nondividing state. Third, since T lymphocytes are in fresh complete medium, quiescent histone synthesis need not be only the result of depletion of nutrients from the growth medium.

2.3.   Stability of Basal and Quiescent Histones

Since both basal and quiescent histones seem to be synthesized in the absence of DNA syntheis, questions arise in terms of their stability and whether they are stably incorporated into chromatin. Pulse–chase experiments indicate that histones labeled during a 1-hr pulse in G_1 (R. S. Wu and Bonner, 1981) or a 2-hr pulse during quiescence (G_0) (R. S. Wu *et al.*, 1982b) were found at the same level in the nuclei two cell generations later for G_1 cells and 7 days later for G_0 cells (R. S. Wu *et al.*, 1983b). In cells labeled during G_1, three nonhistone proteins were found to have half-lives of less than one cell generation, while the basal histones had half-lives of at least five cell generations (R. S. Wu and Bonner, 1981). Both types of histone biosynthesis were unaffected by hydroxyurea treatment (Fig. 6F) (R. S. Wu and Bonner, 1981). UV treatment to stimulate repair DNA synthesis had no effect on quiescent histone synthesis (R. S. Wu *et al.*, 1983a). Furthermore, newly synthesized basal variants are incorporated into the mononucleosome fraction of chromatin within a 1-hr chase during G_1 (R. S. Wu *et al.*, 1983b). Such results suggest that histones synthesized during both G_1 and G_0 are stably incorporated into chromatin independent of DNA synthesis.

3. IMPLICATIONS FOR GENE REGULATION AND CONTROL OF CELL PROLIFERATION

The results presented represent a distinct biochemical difference between G_0 and G_1 states of cells. The ratio of homeomorphous H2A-variant synthesis to heteromorphous variant synthesis (H2A.1 + .2/H2A.X + .Z), in conjunction with the type of H3 variant synthesized, may be useful as biochemical marker(s) for determining the physiological states of cells.

Even more important, the histone-protein synthesis patterns are reflected in messenger RNA (mRNA) levels (Melli *et al.*, 1977; Stein *et al.*, 1978; Stein and Stein, 1980; Marashi *et al.*, 1982) and hence gene activity. Melli *et al.* (1977) and Stein *et al.* (1978) both measured levels of cytoplasmic histone mRNA through the cell cycle. Both reported a large increase in histone mRNA as cells progressed from G_1 to S phase. Melli and co-workers did report G_1 histone mRNA levels 5–7% of those in S phase, levels sufficient to explain the amount of basal histone synthesis. Therefore, we can study a set of specific genes regulated in conjunction with cell state and possibly gain useful insight into the mechanism of control of cell proliferation.

4. CHROMATIN STRUCTURE

In the remaining sections of this chapter, we will discuss chromatin structure, histone genes, and histone mRNAs from the viewpoint of histone variants.

Chromatin in mammalian cells is packaged into an ordered structure (for reviews, see McGhee and Felsenfeld, 1980; Igo-Kemenes *et al.*, 1982). The basic subunit is called a nucleosome, consisting of a piece of DNA [about 200 base pairs (bp)] wrapped around a protein core formed by four of the five histone species. This protein core particle is an octamer composed of two molecules each of H2A, H2B, H3, and H4. Under the electron microscope, each nucleosome resembles a bead with a short piece of DNA protruding at each end. The chromatin in the cell is composed of many beadlike structures interconnected by variable lengths of DNA to form a long string.

The DNA between the nucleosomes has been termed the "linker" or "spacer" DNA, and the fifth histone, H1, is associated with it. The chromatin, after micrococcal-nuclease digestion, shows on gels a fundamental repeat pattern that can differ from tissue to tissue and from species to species. This difference is due to the fact that the "linker" DNA, which is preferentially digested by micrococcal nuclease, can vary in length from tissue to tissue or species to species. Prolonged incubation with micrococcal nuclease results in the complete digestion of "linker" DNA and produces a stable nucleosome core particle composed of a DNA segment 140 bp in length wrapped around the histone octamer. The length of this 140-bp DNA segment is known to be invariant in all chromatin studied so far.

The strings of nucleosomes are packaged into higher-ordered structures by the folding of DNA into the extremely compact form found in the cell nucleus. This second level of chromatin condensation is represented by 250 Å fibers. Various models (for reviews, see Felsenfeld, 1978; McGhee and Felsenfeld, 1980; Igo-Kemenes et al., 1982) for the structure of the 250 Å fibers have been proposed based on electron microscopy, neutron diffraction, nuclease-digestion studies, and electric dichroism data. Histone H1 and the proper ionic environment are required for the formation of these fibers (Labhart et al., 1982). The 250 Å fibers in turn are folded into loops or domains (Caron and Thomas, 1981; Lasters et al., 1981). The loops seem to be attached to a nuclear envelope matrix or scaffold (Berezeny and Coffey, 1974; Aaronson and Blobel, 1975; Adolph et al., 1977). Specific DNA sequences associated with specific proteins may be involved at the base of these loops (Paulson and Lammeli, 1977; Capco et al., 1982).

The discovery of histone variants that are stably incorporated into nucleosomes makes the existence of many possible types of nucleosomes with different core histone composition and different DNA content a reality. Furthermore, the timing of the synthesis of H3.3, H2A.X, and H2A.Z variants relative to the S phase variants may lead to nonrandom localization of specific types of core particles in chromatin, thereby producing nucleosome heterogeneity with functional consequences.

4.1. Functional Significance of Histone Variants

Two opposing positions can be taken on the question of functional significance of histone variants. One extreme position is that variants are merely the allowable diversity in sequence without affecting the same set of functions.

The opposite position is that variants have significantly different functions, although these different functions have not been discovered. Experimental evidence suggests that both positions may be correct. For example, in yeast, there are two H2B variants. Mutants can be made that have only one of the two H2Bs, and the mutants survive (Rykowski *et al.*, 1981). Also, the H2B.2 found in mouse seems to be absent in most other mammals. These two examples provide evidence for the first position.

On the other hand, there is some suggestive evidence for functional differentiation. Simpson (1981) showed that chromatin from early and late sea urchin embryos had different nuclease-digestion and thermal parameters. Blankstein and Levy (1976) showed that the H2A.2/H2A.1 ratio of different Friend erythroleukemia cell lines correlated with the dimethylsulfoxide inducibility of those lines.

In the case of H2As, the differences in sequence between the variants can be quite small or rather large within the same species. H2A.1 and .2 differ from each other by only four positions, whereas H2A.X differs from H2A.1 by somewhat more and H2A.Z by many more. Could H2A.Z have a different function than the other H2A variants because of the very large differences in sequence from H2A.1? Several data support a differentiated role for H2A.Z. First, H2A.Z, as discussed previously, is highly conserved evolutionarily. If a variant did have a critical function, one might expect this variant to have low divergence frequency and be present in all cells of all species. H2A.Z is found in every cell type or species that has been analyzed. Furthermore, it is always present in a rather small but constant fraction (5–10%) of total H2A in all cell types and species. Since approximately 5% of the total nucleosome population could contain H2A.Z, a limited number of strategically placed H2A.Z-containing nucleosomes could be involved in regulating gene activity.

4.2. Nucleosome/Histone Turnover

Quiescent and basal histone synthesis opens the possibility that nucleosomes or histones turn over. If there is no net DNA synthesis during G_1 and G_0, the newly synthesized histones would have to displace existing histones. There could be a shortening of nucleosome repeat to accommodate more histone core particles during G_1 in the case of basal histone synthesis, but this by itself could not accommodate continuous long-term quiescent histone synthesis. If newly synthesized H3.3 displaced only other H3.3 molecules, one would see only H3.3 and not H3.1 and H3.2 turnover. On the other hand, if newly synthesized H3.3 displaced the H3s at random, then one would expect to see a gradual change in the H3 mass pattern of quiescent cells. There is some evidence for the latter. Many mouse tissues such as liver (Bonner *et al.*, 1980), kidney, and brain (Zweidler, 1976) have H3.3 as the major H3 variant, while most tissue-culture lines have little H3.3. Zweidler (1980) also reported that the mass fraction of H3 in H3.3 increased from 15% in livers of newborn mice to 55% in livers of adult mice. This shift is consistent with the result that H3.3 is the only H3 synthesized in quiescent cells. The H3-variant shift in tissues could also be due to turnover of whole cells as well as to nucleosome turnover. Commerford *et al.* (1982) investigated this problem with a pulse–chase experiment in mice and found that the half-life of liver histone was 117 days as compared to 318 days for liver DNA. These results taken together strongly suggest that histone turnover as opposed to cell turnover accounts for at least part of the change in histone-variant ratios. In the case of H1, there is direct evidence that certain H1 variants turn over rather rapidly as cells become quiescent (Sizemore and Cole, 1981; Pehrson and Cole, 1982; Lennox *et al.*, 1982; D'Anna *et al.*, 1982).

Could remodeling of chromatin for transcription also account for some of the nucleosome/histone turnover? Simpson *et al.* (1980) have proposed a model for transcription of nucleosome-associated genes in which nucleosomes dissociate from the DNA as the RNA polymerase reads through, then reassociate with the DNA. Quiescent and basal histone synthesis opens the possibility that the dissociated histones are simply degraded, with newly synthesized histones replacing them in the nucleosome. During such a process, the chromatin could be put into various active or inactive configurations.

4.3. Chromatin from Specific Families of Sequences

The DNA in eukaryotic cells is highly complex. Data obtained by renaturation kinetic studies indicate that the DNA in the eukaryotic genome can be classified into three broad categories: highly repetitive, moderately repetitive, and unique sequences (Britten and Kohne, 1968). These sequences have been further defined cytologically and biochemically (for reviews, see Appels and Peacock, 1978; John and Miklos, 1979; Brutlag, 1980; Jelinek and Schmid, 1982).

The highly repetitive sequences are represented by clustered repeats of short, simple sequences. These sequences may be repeated as many as 10^6–10^7 times and are usually found in centromeric and telomeric heterochromatin. A typical histone composition in chromatin containing highly repetitive DNA was observed by Omori *et al.* (1980) using 1D SDS gels and Zhang and Horz (1982) using both SDS and 1D acetic acid–urea–Triton gels. No significant differences were detected in the composition of minor histone variants between bulk chromatin and chromatin containing highly repetitive sequences.

The moderately repeated sequences can be divided into two categories, short and long. These short and long repeat sequences are usually interspersed with unique sequences. A typical example of the short dispersed repeat sequence is the *Alu* family of repeats found in human and rodent cells (for a review, see Schmid and Jelinek, 1982). The long dispersed repeats resemble sequences in proretroviruses and are best characterized in *Drosophila* and yeast. A common characteristic of all the moderately repeated sequences is that they are highly mobile elements and tend to be transposed to different regions of the genome. Chromatin containing moderately repeated sequences has not been analyzed for histone content.

The unique sequences are represented in DNA approximately one copy per haploid genome. The closest effort in studying the protein composition of the nucleosomes found in a specific region of unique sequence is the work reported by Weisbrod and Weintraub (1981) and Weisbrod (1982). They studied the composition of transcriptionally active chromatin isolated from 14-day-old chicken red blood cells after digestion with nucleases using an "HMG" column (HMG 14 and 17 coupled to agarose or glass beads), which has an affinity for transcriptionally active nucleosomes containing the hemoglobin gene. Using a 1D acid–urea–Triton gel for analysis, the histone core particles found associated with the transcriptionally active hemoglobin gene were shown to have the standard octameric composition of bulk chromatin. However, the minor histone variants, particularly H2A.Z, were not resolved; therefore, the amount of minor variants relative to the amount of the major variants could not be compared. The selectivity of deoxyribonuclease I, an enzyme that preferentially digests transcriptionally active chromatin (Weintraub and Groudine, 1976; Flint and Weintraub, 1977; C. Wu *et al.*, 1979a,b; Stalder *et al.*, 1980; Storb *et al.*, 1981), has been attributed to selective changes in higher-order structures of chromatin, modification or removal of nucleosomes in the region of active genes, or modification of DNA bases through methylation (Groudine and Weintraub, 1981; Weintraub *et al.*, 1981). Perhaps the placement of a minor histone variant such as H2A.Z in or near a specific gene could be involved in differences between active and inactive genes.

The analyses of histone-variant composition in chromatin containing specific sequences of DNA can be summarized as follows: The experiments with SDS–polyacrylamide gel electrophoresis could not resolve the variants. The experiments with 1D acetic acid–urea–Triton gels did not detect any large differences in the major variant composition between bulk chromatin and chromatin containing specific sequences. The content of minor variants, H2A.X, H2A.Z, and H3.3, in chromatin containing specific sequences is less clear because these variants are not easily resolved on 1D gels.

5. HISTONE GENES

The second major area in which studies on histone variants impinge is the area of histone genes. The histone genes (for a review, see Hentchel and Birnsticl, 1981) of sea urchin (Kedes, 1979) and *Drosophila* (Lifton *et al.*, 1977) embryos have been the most studied. Detailed structural maps have revealed that these histone genes belong to the moderately repetitive class of DNA and are found as repetitive and tandemly linked units in the genome with no introns in each of the coding sequences. Noncoding spacer sequences are interdigitated between the coding regions of the five respective histone proteins. These multigenic clusters can be separated as a satellite from bulk chromatin. Since mRNAs isolated from embryos during early development hybridize to these histone genes, they have been classified as belonging to the early-developmental-stage-specific operons. The clustering of tandemly repeated units of early histone genes in the sea urchin embryos may be useful because the rapid rate of cell division may require large amounts of histone proteins as well as very rapid DNA replication (Maxson *et al.*, 1983a). Even though many of the histone gene sequences are known, the linkage map for various histone variants of one class of histone to another class of histone in the multigenic clusters is unknown.

Recent studies on the expression of late-stage-specific histone genes in sea urchins have revealed that the late histone genes may not occur in clusters as tandemly repeating units (Maxson *et al.*, 1983b). The histone genes in many mammals (Heintz *et al.*, 1981; Sittman *et al.*, 1981; Sierra *et al.*, 1982) and birds (Engel *et al.*, 1982; Grandy *et al.*, 1982) may be arranged similarly to the late histone genes of sea urchins.

To date, no genes for the minor histone variants, especially those variants that are synthesized during G_1–G_2, have been isolatd or identified. Since the cloning of most genes involves the specific synthesis of a complementary DNA (cDNA) from a mRNA and since the mRNAs used to clone the histone genes were obtained from log-phase or S-phase cells, the probability of selecting a gene for a basal variant is much less than that of selecting a gene for an S-phase variant.

6. HISTONE MESSENGER RNAs

Two areas of uncertainty concern histone mRNA metabolism: (1) the existence of a polyadenylic acid poly(A) tract on the 3′ terminus of the mRNA and (2) whether histone mRNA transcription and turnover are closely linked to DNA replication. Until recently, the working hypothesis was that the 9 S histone mRNA was devoid of a poly(A) tract at the 3′ terminus, was not present in G_1 cells, and disappeared rapidly when DNA replication was inhibited by inhibitors of DNA synthesis. This general view has not held up when tested experimentally.

Increasing evidence has suggested that at least a portion of some histone mRNAs may be polyadenylated. Such subsets of poly(A)-containing histone mRNAs have been reported in yeasts (Fahrner *et al.*, 1980), HeLa cells (Borun *et al.*, 1977), amphibians (Levenson and Marcu, 1976), sea urchins (Ruderman and Pardue, 1978), and clams (Gabrielli and Baglioni, 1975). Again, the specific histone variants that are represented by the poly(A)-containing mRNAs have not been identified. Furthermore, these experiments were all performed with log-phase or S-phase cells to enhance the yield of histone mRNA. Could the histone mRNAs isolated from G_0 or G_1 cells contain poly(A)? Claims of poly(A) tails in histone mRNA must be treated with some caution. For example, Engel *et al.* (1982) recovered an H3 mRNA in the poly(A)-containing fraction, but this affinity was due to a transcribed A-rich sequence ($A_{11}GAg$) and was not from post-transcriptionally added poly(A).

There are also recent conflicting data on the tight coupling of histone synthesis to DNA synthesis. Groppi and Coffino (1980) have reported that histones may be synthesized at equivalent rates in G_1 and S phases or cycling cells, thus implying the existence of histone mRNAs in G_1 phase of the cell cycle. Marashi *et al.* (1982), using cDNA probes made from S-phase mRNAs, reported that newly synthesized histones were detected only in S-phase cells, and histone mRNAs were not detectable by hybridization in G_1. Furthermore, R. S. Wu and Bonner (1981) have shown that histone synthesis in cycling cells can be separated into basal-type synthesis and S-phase-type synthesis. Basal histone synthesis was less than 10% of the S-phase synthesis but persisted throughout the cell cycle, while S-phase synthesis role in parallel with the increase in DNA synthesis. In view of these findings, the report by Lichtler *et al.* (1982) that H4 mRNA can be separated into six or seven different species is interesting. Could some of these H4 mRNAs be of the basal type? Since H4 resolves into only a single variant on 2D gels but is synthesized during G_1–G_2 as well as S, it is probable that the products of different H4 genes can be detected only at the mRNA level.

Limited sequence data are available on histone mRNAs. The nucleotide sequence for one of the H4 histone mRNAs of the sea urchins *S. purpuratus* and *L. pictus* have been obtained (Grunstein *et al.*, 1973, 1976; Grunstein and Schedl, 1976; Grunstein and Grunstein, 1977). Lichtler *et al.* (1982) have fingerprinted several of the different HeLa cell H4 mRNAs; complete sequence data are not available at present. The existence of noncoding sequences in histone mRNAs has been shown by heteroduplex mapping. Both leader and trailer sequences exist (M. Wu *et al.*, 1976; Holmes *et al.*, 1977; Grunstein and Grunstein, 1977). The 5′ of the histone mRNAs, like that of many eukaryotic mRNAs, is capped with the structure $m^7G(5')pppX^mpY$ (Surrey and Nemer, 1976). Much work remains to be done in correlating and identifying the different mRNAs found for each of the five histone proteins with the different protein variants in each of these classes.

7. *IN VITRO* TRANSCRIPTION

Sea urchin nuclei have been used to transcribe histone mRNA *in vitro* (Shutt and Kedes, 1974; Levy *et al.*, 1978). The *in vitro* transcripts are not polycistronic in size, but range from 100 to 1000 nucleotides in length. The largest transcripts made are bigger than the 9 S cytoplasmic histone mRNA. If the fidelity of *in vitro* transcription corresponds to that of *in vivo* transcription, then posttranscriptional processing of nuclear RNAs must occur in order to generate the shorter polysomal mRNAs. An active processing mechanism may indeed be involved in view of the recent finding that a histone *H3* gene containing introns bounded by the conventional (5′-GT, 3′-AG) splice junctions at the intron–exon borders has been isolated (Engel *et al.*, 1982).

Many laboratories are using the histone genes as a model system to study factors and regulatory DNA sequences that regulate transcription and to examine the fidelity of the *in vitro* transcription process. Grosschedl and Birnstiel (1980) cloned a 6-kilobase sea urchin histone DNA containing all five histone genes alternating with spacer DNA. They studied the expression (transcription) of this cloned DNA in the oocytes of *Xenopus*. They found that predominately *H2A* and *H2B* sequences were transcribed. With the help of cloned deletion mutations upstream from the *H2A* gene, they concluded that there are enhancer/modulating sequences located 5' upstream from the *H2A* genes that regulate the fidelity of transcription and also the actual amount of mRNA synthesized.

8. FUTURE STUDIES

This brief overview of histone-variant studies in the area of chromatin structure and the regulation of histone gene expression points to many directions for future research. Some of the areas will require the development of new technology, and others are currently being actively pursued. For example, the lack of a technique for the separation of nucleosomes based on histone-variant content or DNA sequence makes it difficult to study the arrangement and localization of types of nucleosomes in chromatin. Could particular DNA sequences contain more basal-type variants? How do cells remodel chromatin during transcription or during DNA repair?

The isolation and cloning of a specific histone variant gene would allow studies on histone-variant gene regulation and its involvement in cell behavior. How are the changes in variant synthesis pattern at different physiological states and around the cell cycle reflected at the mRNA level? How does the cell regulate the close to 1:1:1:1 ratio of synthesis of the core histones when some of them have variants? Do the genes for the different histone variants have different promotors? What factors influence the transcriptional activity of the different variant genes? Can or do specific histone variants or core particles influence transcriptional activity? When some of these questions are answered, we will have a better understanding of the biological functions of histone proteins.

REFERENCES

Aaronson, R. P., and Blobel, G., 1975, Isolation of nuclear pore complexes in association with a lamina, *Proc. Natl. Acad. Sci. U.S.A.* **72:**1007–1111.

Abercrombie, M., and Ambrose, E. J., 1962, The surface properties of cancer cells: A review, *Cancer Res.* **22:**525–548.

Adolph, K. W., Cheng, S. M., and Laemmli, U. K., 1977, Role of nonhistone proteins in metaphase chromosome structure, *Cell* **12:**805–816.

Alfageme, C. R., Zweidler, A., Mahowald, A., and Cohen, L. H., 1974, Histones of *Drosophila* embryos, *J. Biol. Chem.* **249:**3729–3736.

Allis, C. D., Glover, C. V. C., Bowen, J. K., and Gorovsky, M. A., 1980, Histone variants specific to the transcriptionally active, amitotically dividing macronucleus of the unicellular eucaryote, *Tetrahymena thermophila, Cell* **20:**609–617.

Appels, R., and Peacock, W. J., 1978, The arrangement and evolution of highly repeated (satellite) DNA sequences with special reference to *Drosophila, Int. Rev. Cytol.* **8:**69–126 (Suppl.).

Ball, D. J., Slaughter, C. A., Hensley, P., and Garrard, W. T., 1983, Amino acid sequence of N-terminal domain of calf thymus histone H2A.Z, *FEBS Lett.* **154:**166–170.

Berezney, R., and Coffey, D. S., 1974, Identification of a nuclear protein matrix, *Biochem. Biophys. Res. Commun.* **60:**1410–1417.

Blankstein, L. A., and Levy, S. B., 1976, Changes in histone f2a2 associated with proliferation of Friend leukaemic cells, *Nature (London)* **260:**638–640.

Bonner, W. M., West, M. H. P., and Stedman, J. D., 1980, Two-dimensional gel analysis of histones in acid extracts of nuclei, cells, and tissues, *Eur. J. Biochem.* **109:**17–23.

Borun, T. W., Ajiro, K., Zweidler, A., Dolby, T. W., and Stephens, R. E., 1977, Studies of human histone messenger RNA II: The resolution of fractions containing individual human histone messenger RNA species, *J. Biol. Chem.* **252:**173–180.

Bradbury, E. M., Inglis, R. J., Matthews, H. R., and Sarner, N., 1973, Phosphorylation of very lysine-rich histones in *Physarum polycephalum:* Correlation with chromosome condensation, *Eur. J. Biochem.* **33:**131–139.

Britten, R. J., and Kohne, D. E., 1968, Repeated sequences in DNA, *Science* **161:**529–540.

Brutlag, D., 1980, Molecular arrangement and evolution of heterochromatic DNA, *Annu. Rev. Genet.* **14:**121–144.

Capco, D. G., Wan, K. M., and Penman, S., 1982, The nuclear matrix: Three-dimensional architecture and protein composition, *Cell* **29:**847–858.

Caron, F., and Thomas, J. O., 1981, Exchange of histone H1 between segments of chromatin, *J. Mol. Biol.* **146:**513–537.

Commerford, S. L., Carsten, A. L., and Cronkite, E. P., 1982, Histone turnover within non-proliferating cells, *Proc. Natl. Acad. Sci. U.S.A.* **79:**1163–1165.

D'Anna, J. A., Jr., and Isenberg, I., 1974, Conformational changes of histone LAK(f2a2), *Biochemistry* **13:**2098–2104.

D'Anna, J. A., Gurley, L. R., and Tobey, R. A., 1982, Syntheses and modulations in the chromatin contents of histones $H1^0$ and H1 during G_1 and S phases in Chinese hamster cells, *Biochemistry* **21:**3991–4000.

Delegeane, A. M., and Lee, A. S., 1981, Coupling of histone and DNA synthesis in the somatic cell cycle, *Science* **215:**79–81.

Elgin, S. C. R., and Weintraub, H., 1975, Chromosomal proteins and chromatin structure, *Annu. Rev. Biochem.* **44:**725–774.

Engel, J. D., Sugarman, B. J., and Dodgson, J. B., 1982, A chicken histone H3 gene contains intervening sequences, *Nature (London)* **297:**434–436.

Fahrner, K., Yarger, J., and Hereford, L., 1980, Yeast histone mRNA is polyadenylated, *Nucleic Acids Res.* **8:**5725–5737.

Felsenfeld, G., 1978, Chromatin, *Nature (London)* **271:**115–122.

Flint, S. J., and Weintraub, H. M., 1977, An altered subunit configuration associated with the actively transcribed DNA of integrated adenovirus genes, *Cell* **12:**783–794.

Franklin, S. G., and Zweidler, A., 1977, Non-allelic variants of histones 2a, 2b, and 3 in mammals, *Nature (London)* **266:**273–275.

Gabrielli, F., and Baglioni, C., 1975, Maternal messenger RNA and histone synthesis in embryos of the surf clam *Spisula solidissima, Dev. Biol.* **43:**254–263.

Grandy, D. K., Engel, J. D., and Dodgson, J. B., 1982, Complete nucleotide sequence of a chicken H2b histone gene, *J. Biol. Chem.* **257:**8577–8583.

Groppi, V. E., Jr., and Coffino, P., 1980, G_1 and S phase mammalian cells synthesize histones at equivalent rates, *Cell* **21:**195–204.

Grosschedl, R., and Birnstiel, M. L., 1980, Spacer DNA sequences upstream of the T-A-T-A-A-A-T-A sequence are essential for promotion of H2A histone gene transcription. *Proc. Natl. Acad. Sci. U.S.A.* **77:**7102–7106.

Groudine, M., and Weintraub, H., 1981, Activation of globin genes during chicken development, *Cell* **24:**393–401.

Grunstein, M., and Grunstein, J., 1977, The histone H4 gene of *Strongylocentrotus purpuratus:* DNA and mRNA sequences at the 5' end, *Cold Spring Harbor Symp. Quant. Biol.* **42:**1083–1092.

Grunstein, M., and Schedl, P., 1976, Isolation and sequence analysis of sea urchin *(Lytechinus pictus)* histone H4 messenger RNA, *J. Mol. Biol.* **104:**323–349.

Grunstein, M., Levy, S., Schedl, P., and Kedes, L. H., 1973, Messenger RNAs for individual histone proteins: Fingerprint analysis and *in vitro* translation, *Cold Spring Harbor Symp. Quant. Biol.* **38:**717–724.

Grunstein, M., Schedl, P., and Kedes, L. H., 1976, Sequence analysis and evolution of sea urchin (*Lytechinus pictus* and *Strongylocentrotus purpuratus*) histone H4 messenger RNAs, *J. Mol. Biol.* **104:**351–369.

Gurley, L. R., Walters, R. A., and Tobey, R. A., 1975, Sequential phosphorylation of histone subfractions in the Chinese hamster cell cycle, *J. Biol. Chem.* **250:**3936–3944.

Gurley, L. R., Walters, R. A., Barham, S. S., and Deaven, L. L., 1978, Heterochromatin and histone phosphorylation, *Exp. Cell Res.* **111:**373–383.

Heintz, N., Zernik, M., and Roeder, R. G., 1981, The structure of the human histone genes: Clustered but not tandemly repeated, *Cell* **24:**661–668.

Hentchel, C., and Birnstiel, M. L., 1981, The organization and expression of histone gene families, *Cell* **25:**301–313.

Hnilica, L. S., 1964, The specificity of histones in chicken erythrocytes, *Experientia* **20:**13–14.

Holley, R. W., 1975, Control of growth of mammalian cells in cell culture, *Nature (London)* **258:**487–490.

Holmes, D. S., Cohen, R. H., Kedes, L. H., and Davidson, N., 1977, Position of sea urchin *(Strongylocentrotus purpuratus)* histone genes relative to restriction endonuclease sites on the chimeric plasmids pSp2 and pSp17, *Biochemistry* **16:**1504–1512.

Igo-Kemenes, T., Horz, W., and Zachau, H., 1982, Chromatin, *Annu. Rev. Biochem.* **51:**89–121.

Isenberg, I., 1979, Histones, *Annu. Rev. Biochem.* **48:**159–191.

Jelinek, W. R., and Schmid, C. W., 1982, Repetitive sequences in eucaryotic DNA and their expression, *Annu. Rev. Biochem.* **51:**813–844.

John, B., and Miklos, G. L. G., 1979, Functional aspects of satellite DNA and heterochromatin, *Int. Rev. Cytol.* **58:**1–114.

Kedes, L. H., 1979, Histone genes and histone messengers, *Annu. Rev. Biochem.* **48:**837–870.

Klotz, I. M., Darnall, D. W., and Langerman, N. R., 1975, Quaternary structure of proteins, in: *The Proteins,* Vol. 1 (H. Neurath and R. L. Hill eds.), Academic Press, New York, pp. 293–411.

Labhart, P., Koller, T., and Wunderli, H., 1982, Involvement of higher order chromatin structures in metaphase chromosome organization, *Cell* **30:**115–121.

Laine, B., Sautiere, P., and Biserte, G., 1976, Primary structure and microheterogeneities of rat chloroleukemia histone H2A (histone HLK, IIbl, or F2a2), *Biochemistry* **15:**1640–1645.

Lasters, I., Muyldermans, S., Wyns, L. and Hamers, R., 1981, Differences in rearrangement of H1 and H5 in chicken erythrocyte chromatin, *Biochemistry* **20:**1104–1110.

Lennox, R. W., Oshima, R. G., and Cohen, L. H., 1982, The H1 histones and their interphase phosphorylated states in differentiated and undifferentiated cell lines derived from murine teratocarcinomas, *J. Biol. Chem.* **257:**5183–5189.

Levenson, R., and Marcu, K., 1976, On the existence of polyadenylated histone mRNA in *Xenopus laevis* oocytes, *Cell* **9:**311–322.

Levy, S., Childs, G., and Kedes, L. H., 1978, Sea urchin nuclei use RNA polymerase II to transcribe discrete histone RNAs larger than messengers, *Cell* **15:**151–162.

Lichtler, A. C., Sierra, F., Clark, S., Wells, J. R. E., Stein, J. L., and Stein, G. S., 1982, Multiple H4 histone mRNAs of HeLa cells are encoded in different genes, *Nature (London)* **298:**195–198.

Lifton, R. P., Goldberg, M. L., Karp, R. W., and Hogness, D. S., 1977, The organization of histone genes in *Drosophila melanogaster:* Functional and evolutionary implications, *Cold Spring Harbor Symp. Quant. Biol.* **42:**1047–1051.

Marashi, F., Baumbach, L., Rickles, R., Sierra, F., Stein, J. L., and Stein, G. S., 1982, Histone proteins in HeLa S_3 cells are synthesized in a cell cycle specific manner, *Science* **215:**683–685.

Mardian, J. K. W., and Isenberg, I., 1978, Yeast inner histones and the evolutionary conservation of histone–histone interactions, *Biochemistry* **17:**3825–3833.

Marzluff, W. F., Jr., Sanders, L. A., Miller, D. M., and McCarty, K. S., 1972, Two chemically and metabolically distinct forms of calf thymus histone F3, *J. Biol. Chem.* **247:**2026–2033.

Matsui, S. I., Seon, B. K., and Sandberg, A. A., 1979, Disappearance of a structural chromatin protein A24 in mitosis: Implications for molecular basis of chromatin condensation, *Proc. Natl. Acad. Sci. U.S.A.* **76:**6386–6390.

Maxson, R., Mohun, T., and Kedes, L., 1983a, Expression of specific genes: Histone genes, in: *Eucaryotic Genes: Their Structure, Activity and Regulations* (N. McLean, S. Gregory, and R. Flavell, eds.), Butterworth, London (in press).

Maxson, R., Mohun, T., Gormezano, G., Childs, G., and Kedes, L. H., 1983b, Distinct organization and patterns of expression of early and late histone gene sets in the sea urchin, *Strongylocentrotus purpuratus, Nature* **301:**120–125.

McGhee, J. D., and Felsenfeld, G., 1980, Nucleosome structure, *Ann. Rev. Biochem.* **49:**1115–1156.

Melli, M., Spinelli, G., and Arnold, E., 1977, Synthesis of histone messenger RNA of HeLa cells during the cell cycle, *Cell* **12:**167–174.

Neelin, J. M., Callahan, P. X., Lamb, D. C., and Murray, K., 1964, The histones of chicken erythrocyte nuclei, *Can. J. Biochem.* **42:**1743–1752.

Newrock, K. M., Alfageme, C. R., Nardi, R. V., and Cohen, L. H., 1978, Histone changes during chromatin remodeling in embryogenesis, *Cold Spring Harbor Symp. Quant. Biol.* **42:**421–431.

Omori, A., Igo-Kemenes, T., and Zachau, H. G., 1980, Different repeat lengths in rat satellite I DNA containing chromatin and bulk chromatin, *Nucleic Acids Res.* **8:**5363–5375.

Palmer, D., Snyder, L. A., and Blumenfeld, M., 1980, *Drosophila* nucleosomes contain an unusual histone like protein, *Proc. Natl. Acad. Sci. U.S.A.* **77:**2671–2675.

Pantazis, P., and Bonner, W. M., 1981, Quantitative determination of histone modification, *J. Biol. Chem.* **256:**4669–4675.

Paulson, J. R., and Laemmli, U. K., 1977, The structure of histone depleted metaphase chromosomes, *Cell* **12:**817–828.

Pehrson, J. R., and Cole, R. D., 1982, Histone H1 subfractions and $H1^0$ turnover at different rates in nondividing cells, *Biochemistry* **21:**456–460.

Ruderman, J. V., and Pardue, M. L., 1978, A portion of all major classes of histone messenger RNA in amphibian oocytes is polyadenylated, *J. Biol. Chem.* **253:**2018–2025.

Rykowski, M., Wallis, J., Choe, J., and Grunstein, M., 1981, Histone 2B sub-types are dispensible during the yeast cell cycle, *Cell* **25:**477–487.

Sautiere, P., Wonters-Tyrou, D., Laine, B., and Biserte, G., 1975, Structures of histone H2A (histone ALK, IIb1 or F2a2), in: *The Structure and Function of Chromatin: a Ciba Foundation Symposium,* Vol. 28, Elsevier/North-Holland, Amsterdam, pp. 77–88.

Schmid, C. W., and Jelinek, W. R., 1982, The Alu family of dispersed repetitive sequences, *Science* **216**:1065–1070.

Shutt, R., and Kedes, L. H., 1974, Synthesis of histone mRNA sequences in isolated nuclei of cleavage stage sea urchin embryos, *Cell* **3**:283–292.

Sierra, F., Lichtler, A., Marashi, F., Rickles, R., Van Dyke, T., Clark, S., Wells, J., Stein, G., and Stein, J., 1982, Organization of human histone genes, *Proc. Natl. Acad. Sci. U.S.A.* **79**:1795–1799.

Simpson, R. T., 1981, Modulation of nucleosome structure by histone subtypes in sea urchin embryos, *Proc. Natl. Acad. Sci. U.S.A.* **78**:6803–6807.

Simpson, R. T., Stein, A., Bitter, G. A., and Kunzler, P., 1980, Chromatin structure and function: A model for transcription of nucleosome associated DNA, in: *Novel ADP-Ribosylation of Regulatory Enzymes and Proteins* (M. E. Smulson and T. Sugimura, eds.), Elsevier/North-Holland, New York, pp. 133–142.

Sittman, D. B., Chiu, I. M., Pan, C. J., Cohn, R. H., Kedes, L., and Marzluff, W. F., 1981, Isolation of two clusters of mouse histone genes, *Proc. Natl. Acad. Sci. U.S.A.* **78**:4078–4082.

Sizemore, S. R., and Cole, R. D., 1981, Asynchronous appearance of newly synthesized histone H1 subfractions in HeLa chromatin, *J. Biol. Chem.* **90**:415–417.

Smith, B. J., and Johns, E. W., 1980, Isolation and characterization of subfractions of nuclear protein H1^0, *FEBS Lett.* **110**:25–29.

Spiker, S., and Isenberg, I., 1977a, Evolutionary conservation of histone–histone binding sites: Evidence from interkingdom complex formation, *Cold Spring Harbor Symp. Quant. Biol.* **42**:157–163.

Spiker, S., and Isenberg, I., 1977b, Cross-complexing pattern of plant histones, *Biochemistry* **16**:1819–1826.

Stalder, J., Groudine, M., Dodgson, J. B., Engel, J. D., and Weintraub, H., 1980, Hb switching in chickens, *Cell* **19**:973–980.

Stein, G. S., and Stein, J. L., 1980, Regulation of histone gene expression during the cell cycle and coupling of histone gene expression with read out of other genetic sequences, in: *Cell Growth* (C. Nicolini, ed.), Plenum Press, New York, pp. 377–409.

Stein, G. S., Stein, J. L., Park, W. D., Detke, S., Lichtler, A. C., Sheppard, E. A., Jansing, R. L., and Phillips, I. R., 1978, Regulation of histone gene expression in HeLa S$_3$ cells, *Cold Spring Harbor Symp. Quant. Biol.* **42**:1107–1120.

Storb, U., Wilson, R., Selsing, E., and Walfield, A., 1981, Rearranged and germline immunoglobulin κ genes: Different states of DNase I sensitivity of constant κ genes in immune competent and non-immune cells, *Biochemistry* **20**:990–996.

Surrey, S., and Nemer, M., 1976, Methylated blocked 5′ terminal sequences of sea urchin embryo messenger RNA classes containing and lacking poly(A), *Cell* **9**:589–595.

Todaro, G. J., Lazar, G. K., and Green, H., 1965, The initiation of cell division in a contact-inhibited mammalian cell line, *J. Cell Comp. Physiol.* **66**:325–334.

Urban, M. K., Franklin, S. G., and Zweidler, A., 1979, Isolation and characterization of the histone variants in chicken erythrocytes, *Biochemistry* **18**:3952–3960.

Von Holt, C., Strickland, W. N., Brandt, W. F., and Strickland, M. S., 1979, More histone structures, *FEBS Lett.* **100**:201–218.

Weintraub, H., and Groudine, M., 1976, Chromosomal subunits in active genes have an altered composition, *Science* **193**:848–856.

Weintraub, H., Larsen, A., and Groudine, M., 1981, α Globin gene switching during the development of chicken embryos: Expression and chromosome structure, *Cell* **24**:333–344.

Weisbrod, S. T., 1982, Properties of active nucleosomes as revealed by HMG 14 and 17 chromatography, *Nucleic Acids Res.* **10**:2017–2041.

Weisbrod, S. T., and Weintraub, H., 1981, Isolation of actively transcribed nucleosomes using immobilized HMG 14 and 17 and an analysis of α globin chromatin, *Cell* **23**:391–400.

West, M. H. P., and Bonner, W. M., 1980a, Histone 2A, a heteromorphous family of eight protein species, *Biochemistry* **19**:3238–3245.

West, M. H. P., and Bonner, W. M., 1980b, Histone 2B can be modified by the attachment of ubiquitin, *Nucleic Acids Res.* **8**:4671–4680.

West, M. H. P., and Bonner, W. M., 1983, Comparison of tryptic peptides of H2A variants from mouse on peptide gels, *Comp. Biochem. Physiol.* (in press).

Wu, C., Bingham, P. M., Livak, K. J., Holmgren, R., and Elgin, S. C. R., 1979a, The chromatin structure of specific genes. I. Evidence for higher order domains of defined DNA sequence, *Cell* **16**:797–806.

Wu, C., Wong, Y. C., and Elgin, S. C. R., 1979b, The chromatin structure of specific genes. II. Disruption of chromatin during gene activity, *Cell* **16**:807–814.

Wu, M., Holmes, D. S., Davidson, N., Cohn, R. H., and Kedes, L. H., 1976, The relative positions of sea urchin histone genes on the chimeric plasmids pSp2 and pSp17 as studied by electron microscopy, *Cell* **9**:163–169.

Wu, R. S., and Bonner, W. M., 1981, Separation of basal histone synthesis from S-phase histone synthesis in dividing cells, *Cell* **27**:321–330.

Wu, R. S., Kohn, K. W., and Bonner, W. M., 1981, Metabolism of ubiquitinated histones, *J. Biol. Chem.* **256**:5916–5920.

Wu, R. S., Nishioka, D., and Bonner, W. M., 1982a, Differential conservation of histone 2A variants between mammals and sea urchins, *J. Cell Biol.* **93**:426–431.

Wu, R. S., Tsai, S., and Bonner, W. M., 1982b, Patterns of histone variant synthesis can distinguish G_0 from G_1 cells, *Cell* **31**:367–374.

Wu, R. S., Tsai, S., and Bonner, W. M., 1983a, Changes in histone H3 composition and synthesis pattern during lymphocyte activation, *Biochemistry* **22**:3868–3873.

Wu, R. S., Perry, L. J., and Bonner, W. M., 1983b, Fate of newly synthesized histones in G_1 and G_0 cells, *FEBS Lett.* (in press).

Zhang, X. Y., and Horz, W., 1982, Analysis of highly purified satellite DNA containing chromatin from the mouse, *Nucleic Acids Res.* **10**:1481–1494.

Zweidler, A., 1976, Complexity and variability of the histone complement, *Life Sci. Res. Rep.* **4**:187–196.

Zweidler, A., 1980, Nonallelic histones in development and differentiation, in: *Gene Families of Collagen and Other Proteins* (D. J. Prockop and P. C. Champe, eds.), Elsevier/North-Holland, Amsterdam and New York, pp. 47–56.

4

Mechanisms for Evolutionary Divergence within the Prolactin Gene Family

NANCY E. COOKE

1. INTRODUCTION

Transcriptional control of the expression of eukaryotic genes encoding proteins appears to be regulated both by DNA structures far removed from the gene (Klar *et al.*, 1981; Nasmyth *et al.*, 1981) and by signals found immediately upstream from the start of transcription (Breathnach and Chambon, 1981). The differential expression of members within a family of genes could therefore be dependent on structural differences adjacent to those genes. The exact location and nature of several of these adjacent regulatory regions have been examined in detail. The TATAA homology that occurs 20–30 base pairs (bp) 5' to the start of transcription appears to specify the nucleotide at which RNA synthesis begins (Corden *et al.*, 1980). A second set of sequences that are necessary for transcriptional efficiency is located upstream from the TATAA homology. The exact site of these modulating regions has recently been determined for the thymidine kinase gene of herpes simplex virus (McKnight and Kingsbury, 1982). In this gene, mutations in a guanine-rich segment between bases -7 and -61 and a cytosine-rich segment between bases -80 and -105 markedly decrease the efficiency of transcription. Although the

NANCY E. COOKE ● Endocrine Section, Department of Medicine, and Department of Human Genetics, University of Pennsylvania, Philadelphia, Pennsylvania 19104.

-61 to -80 region of this gene contains the "CAAT" homology (Benoist *et al.*, 1980; Efstratiadis *et al.*, 1980; Liebhaber *et al.*, 1980) found in a wide variety of genes, it does not appear to affect the level of thymidine kinase gene transcription. Steroid hormones induce transcription in a number of genes. DNA sequences necessary for this induction have been shown to be linked to the 5' flanking region of such glucocorticoid-induced genes (Hynes *et al.*, 1981; Lee *et al.*, 1981; Robins *et al.*, 1982). Further information on specific regulatory structures within other protein-coding genes may be obtained by comparison of the structure of genes within a family that are expressed differentially in response to different signals.

Prolactin (PRL), growth hormone (GH), and chorionic somatomammotropin (CS) are three single-chain polypeptide hormones encoded by genes that have clearly evolved from a common precursor (Niall *et al.*, 1971; Cooke and Baxter, 1982). PRL and GH are each secreted by a specialized set of cells in the anterior pituitary, and their circulating levels are regulated by neurotransmitters from the hypothalamus and by other hormones. CS is produced by the placenta in amounts that increase throughout pregnancy in humans. The level of CS parallels placental mass, and no regulating factor has been found (Daughaday, 1981; Simpson and MacDonald, 1981). An established estrogen-induced rat anterior pituitary cell line differs from normal pituitaries in that the single cell line produces and secretes both PRL (Tashjian *et al.*, 1970) and GH (Yasumura *et al.*, 1966; Tashjian *et al.*, 1968). Expression of these genes is differentially effected by a number of hormonal and other stimuli (Martin and Tashjian, 1977). *In vivo*, there are complex and multiple mechanisms controlling the synthesis and secretion of PRL. However, PRL is mainly under negative dopaminergic regulation (Maurer, 1980; Takahara *et al.*, 1974; Ben-Jonathan *et al.*, 1977), while GH may be predominantly under positive control (Frohman *et al.*, 1971; Krulich *et al.*, 1968). Glucocorticoid hormones induced GH production at the transcriptional level in rat pituitary cells (Tushinski *et al.*, 1977), whereas they do not induce, and in higher concentrations may actually suppress, PRL production both *in vivo* (Nicoll and Meites, 1964) and *in vitro* (Tashjian *et al.*, 1970). The mechanism by which such differential expression has evolved within this gene family is a matter of interest, since understanding this mechanism will contribute to knowledge about the regulation of gene expression. A possible mechanism for rapid divergence of regulatory response among several genes within this family has been delineated through the structural analysis and comparison of the rat *PRL* and rat *GH* genes. Evidence will be presented in this review to suggest that the 5' flanking region and therefore some of the transcription-control regions of this gene family have evolved and diverged by insertion of large fragments of DNA.

2. GENE NUMBER WITHIN THE PROLACTIN FAMILY

PRL is a member of a multigene family that includes *GH* and *CS* and resides on two chromosomes in man. *PRL* is located on chromosome 6 (Owerbach *et al.*, 1981), and the *GH–CS* cluster is located on chromosome 17 (Owerbach *et al.*, 1980). There has been an apparent duplication in the number of *GH–CS* genes as one proceeds up the evolutionary scale. In rats, there is one known *GH* gene (Barta *et al.*, 1981; Page *et al.*, 1981), whereas in primates, there are three or four and in man up to seven genes (Moore *et al.*, 1982). In man, *GH* and *CS* are 85% homologous, 92% at the nucleotide level (Martial *et al.*, 1979; Cooke *et al.*, 1981), while the homology must be much less in rat, in which no gene homologous to *GH* could be isolated by low-stringency hybridization to ^{32}P-labeled rat *GH* complementary DNA (cDNA) (N. Cooke, unpublished data). The homology between *GH* and *CS* suggests that they diverged about 10 million years ago (Cooke *et al.*, 1981), long after the mammalian radiation began and long after the first presumed requirement for placental hormones. A hypothesis to account for their similarity proposes that *GH* and *CS* first diverged at a time more consistent with the emergence of the placenta, but then, perhaps in primate–human evolution, a gene conversion between *GH* and *CS* about 10 million years ago made the two genes more similar (Cooke *et al.*, 1981; Moore *et al.*, 1982). This was then followed by the usual slower rate of divergence to the present.

    PRL and *GH* diverged about 392 million years ago (Cooke *et al.*, 1981). The number of rat *PRL* genes was determined by comparing the restriction map of a cloned and sequenced *PRL* gene to the restriction map of total rat liver DNA, which was Southern-blotted and hybridized with a rat *PRL* cDNA probe (Fig. 1A). The identity of the maps throughout the 11.5 kilobases (kb) of gene sequence (Fig. 1A, B) and into the flanking regions on both sides of the genes suggest that there is only one rat *PRL* gene, unless two or more genes are embedded in identical genomic environments. Consequently, any differential regulatory mechanisms that distinguish between rat *PRL* and *GH* have only one gene template for each hormone with which to interact.

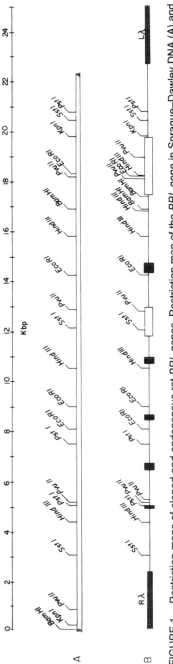

FIGURE 1. Restriction maps of cloned and endogenous rat *PRL* genes. Restriction map of the *PRL* gene in Sprague–Dawley DNA (A) and map of the λCharon 4A clone containing a rat *PRL* recombinant clone (B) are drawn to the kilobase scale indicated at the top and compared. Exons of the cloned gene are drawn as black rectangles, middle repetitive DNA regions as white rectangles. (L, R) Left and right phage arms. From Cooke and Baxter (1982).

3. THE RAT PROLACTIN GENE IS FIVEFOLD LARGER THAN THE RAT GROWTH HORMONE GENE

Rat *GH* (Page *et al.*, 1981; Barta *et al.*, 1981) and rat *PRL* (Maurer *et al.*, 1981; Cooke and Baxter, 1982) have both been sequenced and consist of five exons and four introns. However, rat *GH* is 2.1 kb pairs in length, while rat *PRL* is 11.5 kb pairs, 5 times larger (Fig. 2B). Furthermore, preliminary data suggest that a similar size discrepancy exists between the human *GH* (DeNoto *et al.*, 1981) and human *PRL* genes (N. Cooke, unpublished data). Despite this enormous size discrepancy, the splicing points in the rat *PRL* and rat *GH* genes are at identical sites when the two protein structures are aligned to maximize amino acid homology (Cooke and Baxter, 1982) (Fig. 2A). Conservation of intron location within the coding portion of individual members of gene families is a striking and frequently found phenomenon (Wahli *et al.*, 1980; Efstratiadis *et al.*, 1980). This structural similarity between these two genes substantiates the hypothesis that these two genes evolved from a common ancestor that may have had an exon–intron arrangement similar to those of these two existing genes. One outstanding and unexplained discrepancy exists between *GH* and *PRL* gene structures— the total lack of homology and the difference in size between Exons I in each gene. While the overall structure of the rat *PRL* and rat *GH* genes implied a common precursor, the lack of homology and dissimilarity in length between Exon I of *GH* and *PRL* suggest that the 5′ ends may have had independent origins. A discussion of evidence for such a hypothesis follows.

4. EXONS I OF RAT PROLACTIN AND GROWTH HORMONE ARE FLANKED BY DIRECT REPEATS

The rat *PRL* gene sequence is shown in Fig. 3 (Cooke and Baxter, 1982). Only the sequence in regions contiguous with exons is displayed, including 423 bp of 5′ flanking region and 5069 bp of the gene. This primary sequence was searched by computer analysis for patterns of homology. A direct 10-bp repeat was found that flanks Exon I at bases -77 to -68 and in Intron A at 340–349 (Figs. 3 and 4A). Similarly, a 12-bp repeat (80% homologous) was found flanking Exon I of the rat *GH* gene (Fig. 4A). The *GH* repeats occurred at bases -61 to -50 and 184–196. In both genes, the direct repeats encompass the TATAA region, messenger RNA (mRNA) cap sites, and all

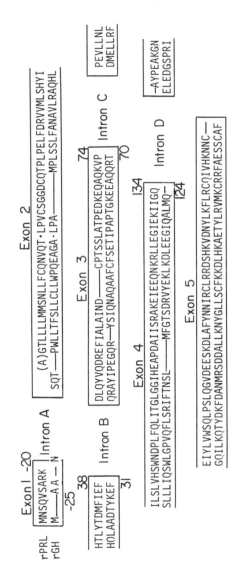

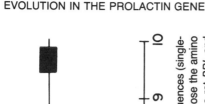

FIGURE 2. Comparison of rat *PRL* and rat *GH* splicing-site locations and overall gene structure. In (A), the amino acid sequences (single-letter code) of rat *pre-PRL* and rat *pre-GH* have been aligned to maximize homology (Cooke *et al.*, 1980). Rectangles enclose the amino acids encoded by each exon. The carboxy-terminal amino acid encoded by each exon is numbered. In (B), the sizes of the rat *PRL* and rat *GH* genes are compared. Exons are indicated by black rectangles, dispersed repetitive DNA by the white and hatched rectangles. From Cooke and Baxter (1982).

```
                          -29
              Met Asn Ser Gln Val Ser Ala Arg Lys
              ATG AAC AGC CAG GTG TCA GCC CGG AAA

                                              -21
                                               -20
              Ala Gly Thr Leu Leu Leu Met Met Ser Asn
              GCA GGG ACA CTC CTC CTG ATG ATG TCA AAC

              Intron A
              825bp

         1
    Leu Leu Phe Cys Gln Asn Val Gln Thr Leu Pro Val Cys Ser Gly Gly Asp Leu Phe Glu Pro Leu Pro Leu Asp His Tyr Ile
    CTT CTG TTC TGC CAA AAT GTG CAG ACC CTG CCA GTC TGT TCT GGT GGC GAC CTG TTT GAC CCT GTG GTC ATG CTT TAC TAC AATC
                                    38
    His Thr Leu Tyr Thr Asp Met Phe Ile Glu Phe
    CAT ACC CTG TAT ACA GAT ATG TTT ATT GAA TTT

              Intron B
              1000bp

                                                         39
    Glu Lys Gln Tyr Val Gln Asp Arg Glu Phe Ile Ala Lys Ala Ile Asn Cys Pro Thr Ser Ser Leu Ala Thr Pro Glu Asp Lys
    GAA AAA CAG TAT GTC CAA GAT CGT GAG TTT ATT GCC AAG GCC ATC AAT GAC TGC CCC ACT TCT TCC CTA GCT ACT CCT GAA GAC AAG

              Intron C
              1750bp

                          74
    Glu Gln Ala Gln Lys Val Pro
    GAA CAA GCC CAG AAA GTC CCT
```

75

Pro Glu Val Leu Leu Asn Leu Ile Ser Leu Val His Ser Trp Asn Asp Pro Leu Ile Thr Gly
134

Leu Gly Ile His Glu Ala Pro Ala Ile Ile Ser Arg Ala Lys Glu Ile Glu Glu Gly Ile Leu Leu Arg Leu Leu Ser Gln

Intron D
1100bp

135
Ala Tyr Pro Glu Ala Lys Gly Asn Glu Ile Tyr Leu Val Trp Ser Val Gln Leu

Pro Ser Leu Gln Gly Val Asp Glu Glu Ser Lys Asp Leu Ala Phe Tyr Asn Asn Ile Arg Cys Leu Arg Arg Asp Ser His Lys Val Asp Ser Tyr Leu Lys Phe Leu
197

Arg Cys Gln Ile Val His Lys Asn Asn Cys OC

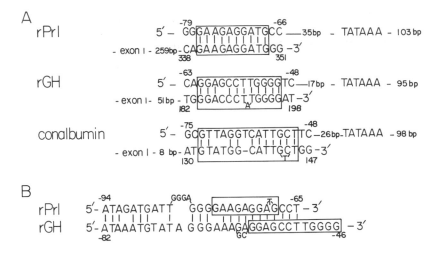

FIGURE 4. Direct repeats flanking Exon I of rat *PRL*, rat *GH,* and conalbumin. (A) Diagram of an isolated segment of the 5′ flanking region, promoter site, first exon, and part of the first intron of three genes. Rectangles surround a decanucleotide repeat in rat *PRL* (Cooke and Baxter, 1982), a 12-bp repeat (one mismatch, one insertion–deletion) in rat *GH* (Barta *et al.,* 1981), and a 14-bp repeat (79% homology, two inser- tions–deletions) in conalbumin (Cochet *et al.,* 1979). (B) An area of homology in the 5′ flanking regions of rat *PRL* and rat *GH*. The 5′ direct repeat in each gene shown in (A) is enclosed by rectangles in (B). Numbering of bases starts at the cap site (+1). From Cooke and Baxter (1982).

of Exon I. Both genes lack CAAT homology regions, which in both cases should lie just upstream from the repeat in the flanking region.

In the rat *PRL* gene, a second AATAAA sequence is located 75 bases upstream from the repeat in the flanking region. This AATAAA is itself preceded 40 bases upstream by CAAAT (see Fig. 3). In the rat *GH* gene, a second AATAAA lies 21 bases 5′ to its flanking repeat, but lacks the CAAT sequence. The finding of these duplicated homology regions, the lack of homology between Exons I of *GH* and *PRL*, and the dissimilarity in size of the coding portions of Exon I suggests that the 426-and 257-bp regions flanked

---

←

FIGURE 3. DNA sequence of the rat *PRL* gene. The numbered amino acids of rat *pre-PRL* are shown. The TATAAA is underlined and the putative cap site is denoted with an asterisk. The decanucleotide repeat flanking Exon I is enclosed in rectangles. The up-stream AATAAA and CAAAT homologies are indicated by arrowheads. The black circles in Intron D indicate the *Bal*I sites shown in Fig. 5, and the underlined sequences cor-respond to L′, L, and R, also shown in Fig. 5. From Cooke and Baxter (1982).

by the direct repeats may have been inserted into the rat *PRL* and rat *GH* genes separately after the time of their divergence. The insertion may have occurred just 3′ to the original "promoter" region of these original genes.

The presence of direct repeats suggests that these sequences may have been inserted by mechanisms proposed for prokaryotic, viral, and eukaryotic insertion sequences and transposons (Calos and Miller, 1980). The insertion of these mobile genetic elements characteristically results in a duplication of host target-site DNA bracketing the inserted DNA. Recently, genomic rearrangements involving extrachromosomal circular duplex DNA have been documented in the human genome (Calabretto *et al.*, 1982) and have implicated *Alu* repeat sequences as a possible vector. Several reported *Alu*-like sequences have been flanked by direct repeats (Barta *et al.*, 1981; Bell *et al.*, 1980). A region of middle repetitive DNA was localized just 5′ to the rat *PRL* gene by mapping the gene with a total genomic probe. Its precise location and sequence are at present unknown. Such a generalized fluidity within the genome no longer seems surprising and most likely represents a common mechanism of rapid evolutionary divergence. Finding direct repeats surrounding Exons I of *PRL* and *GH* is consistent with an insertional event, but only circumstantial evidence that one has occurred.

Prokaryotic transposable elements display a wide spectrum in their selectivity for host target DNA, but in general, most preferentially integrate in certain "hot spots," although integration can occur at other points as well (Calos and Miller, 1980). The 5′ flanking sequences of rat *PRL* and rat *GH* are quite divergent. However, the DNA sequences surrounding the more 5′ of the direct repeats are highly homologous (Fig. 4B). This may represent the remnant of a hot spot for integration, since one must postulate that each gene was the host of a separate insertional event based on the lack of homology between the actual inserted regions.

Since Exon I of rat *PRL* displays 67% (18/27) nucleotide homology to Exon I of human *PRL* (Cooke *et al.*, 1981) and retains the Intron A splicing location (N. Cooke, unpublished results), and the same is true for *GH* [67% (107/160) nucleotide homology in Exons I of rat and man], one must postulate that the insertional event occurred before the divergence of rat and man, during or before the mammalian radiation that occurred 85–100 million years ago (Romero-Herrera *et al.*, 1973; McKenna, 1969). Since the direct repeat is no longer present in human *GH* (although the upstream AATAAA is), we must presume that the preservation of the repeats in the two rat genes might be fortuitous. Alternatively, their preservation may indicate some functional constraint on their structure. In the case of rat *GH*, the 5′ repeat lies in the

−47 to −61 guanine-rich region shown by McKnight and Kingsbury (1982) to regulate the level of transcription in the herpes simplex virus thymidine kinase gene. Mutations in this region of the thymidine kinase gene decreased transcription by 10-fold. The 5′ repeat in rat *PRL* lies between the two thymidine kinase regulatory regions at −80 to 1105 and −47 to −61, and preservation of the spacing between these regions may be important. Also of note, in *PRL*, both these regions are markedly AT-rich in contrast to *GH* and the herpes simplex virus thymidine kinase gene, in which they are markedly GC-rich. Possibly these nucleotide preferences are another structural element contributing to the differential regulation of *GH* and *PRL*. The proposed insertion of Exon I may not be a phenomenon totally unique to the rat *GH* and rat *PRL* genes. For example, a direct repeat flanking Exon I in chick conalbumin (Cochet *et al.*, 1979) was also detected (Fig. 4A).

5.   A SPLICING ANOMALY IN THE RAT PROLACTIN GENE AT THE 3′ END OF INTRON A

In the rat *PRL* gene, the four introns interrupt the cDNA coding regions between codons −20/−19, 38/39, 74/75, and 134/135. The precise positions of the splicing sites were assigned by locating the characteristic GT at the 5′ end of the intron and AG at the 3′ end of the intron (Breathnach *et al.*, 1978), as well as by optimizing homology to U1 RNA (Seif *et al.*, 1979; Rogers and Wall, 1980; Lerner *et al.*, 1980). The actual position of the 3′ splicing junction of Intron A is ambiguous as predicted by Taylor *et al.* (1981). There are two possible splice acceptor sites. One includes an alanine (GCA) in the signal peptide as shown in Fig. 3, and the other excludes this alanine and splices directly to glycine (GGG) at amino acid position −19. The splicing junction that retains the alanine preserves the greatest complementarity with rat U1 RNA (60% vs. 40%). Two separate rat *PRL* cDNAs sequenced from estrogen-stimulated rats contained the alanine codon (Gubbins *et al.*, 1980; Taylor *et al.*, 1981), while a third rat *PRL* cDNA clone isolated from estrogen-stimulated rats that had received hypothalamic ablation did not contain this codon (Cooke *et al.*, 1980). The identification of both mRNA species, although not in one single rat, is consistent with the prediction of alternative splicing within the single rat *PRL* gene. It is possible that this anomalous splicing developed as a result of the postulated insertion of Exon I, although this remains speculative. A similar situation of anomalous splicing does not exist in this portion of the rat *GH* gene.

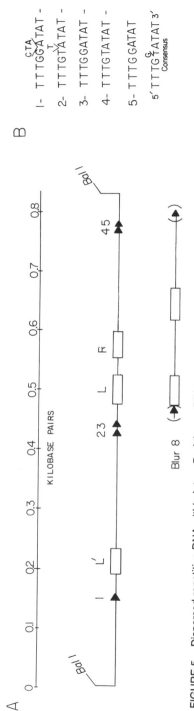

B

1- TTTGGᶜᵀᴬGATAT -

2- TTTGTᵀATAT -

3- TTTGGATAT -

4- TTTGTATAT -

5- TTTGGATAT

5' TTTGᴳᵀTATAT 3'
Consensus

FIGURE 5. Dispersed repetitive DNA within Intron D of the rat *PRL* gene. (A) Diagram of the segment in Intron D of the rat *PRL* gene that hybridizes to ³²P-labeled rat total DNA. (The *Bal*I sites are indicated by black circles in Fig. 3.) Each white rectangle (L', L, and R) contains homology to a 40-bp segment of the human *Alu* family sequence. The actual sequence in each case is underlined in Fig. 3. Blur 8 (Rubin *et al.*, 1980) is a representative member of the human *Alu* family, and its structure is shown for comparison. The arrowheads in parentheses indicate that some *Alu*-like sequences are flanked by direct repeats. The numbered arrowheads represent actual repeats flanking the rat *PRL* Intron D *Alu*-like sequences. The actual sequences of these repeats are shown in (B), along with a derived consensus sequence.

6. DISPERSED REPETITIVE DNA ASSOCIATED WITH THE PROLACTIN GENE FAMILY

Two regions of dispersed repetitive DNA (Jelinek and Schmid, 1982) are present in the cloned 20-kb-pair genomic fragment containing the rat *PRL* gene, one in Intron D and one in the 3' flanking region of the gene (see Fig. 1). The rat *GH* gene contains a segment of repetitive DNA in Intron B (Barta *et al.*, 1981; Page *et al.*, 1981). All three of these repetitive units are unique and do not cross-hybridize. The repetitive element in Intron D, shown between the dots in Fig. 3, appears to contain two *Alu*-like units, the upstream unit having diverged more from the Blur 8 consensus sequence (Rubin *et al.*, 1980) (Fig. 5). Both units are flanked by a set of direct repeats, as is the dispersed repetitive unit in Intron B of the *GH* gene. The 9-bp direct repeats flanking the *PRL* repetitive units are quite similar (Fig. 5B). Insertion and subsequent divergence of these repetitive sequences may serve as a mechanism for rapid expansion of intron DNA. This may account for the generation of some of the size discrepancy between the introns in rat *PRL* and rat *GH* and provides further suggestive evidence for the theory that large insertions have hastened divergence within this gene family. Such insertions may also serve to limit interaction (gene conversion) between the members of the family.

7. SUMMARY

The splicing junctions of Introns A, B, C, and D of the *PRL* and *GH* genes are of the same class and occur at analogous positions despite about 390 million years since their duplication from a common ancestor and their separation onto different chromosomes. During this divergence, the *PRL* gene has accumulated 5-fold more intron DNA than the *GH* gene. Both genes contain dispersed repetitive sequence within introns, flanked by short direct repeats. This suggests an insertional mechanism as one mode of divergence of their introns. Exons I of *PRL* and *GH* are dissimilar in length and non-homologous. Intron A of *PRL* contains an ambiguous splice acceptor. Since the 5' flanking region and Exon I of each gene are flanked by direct repeats that are embedded in homologous regions, we postulate that the areas flanked by repeats were inserted independently into each gene before the divergence of rat and man and may be partially responsible for the differential regulation of gene expression observed for *PRL* and *GH*.

ACKNOWLEDGMENTS. Figures 1, 2, 3, and 4 have been reprinted by permission from *Nature (London)*, copyright 1982, Macmillan Journals, Ltd. I wish to thank Stephen Liebhaber for many helpful discussions.

REFERENCES

Barta, A., Richards, R. I., Baxter, J. D., and Shine, J., 1981, Primary structure and evolution of rat growth hormone gene, *Proc. Natl. Acad. Sci. U.S.A.* **78**:4867–4871.

Bell, G. I., Pictet, R., and Rutter, W. J., 1980, Analysis of the region flanking the insulin gene and sequence of an *Alu* family member, *Nucleic Acids Res.* **8**:4091–4109.

Ben-Jonathan, N., Oliver, C., Weiner, H. J., Mical, R. S., and Porter, J. C., 1977, Dopamine in hypophysial portal plasma of the rat during the estrous cycle and throughout pregnancy, *Endocrinology* **100**:452–458.

Benoist, C., O'Hare, K., Breathnach, R., and Chambon, P., 1980, The ovalbumin gene-sequence of putative control regions, *Nucleic Acids Res.* **8**:127–142.

Breathnach, R., and Chambon, P., 1981, Organization and expression of eukaryotic split genes encoding for proteins, *Annu. Rev. Biochem.* **50**:349–383.

Breathnach, R., Benoist, C., O'Hare, K., Gannon, F., and Chambon, P., 1978, Ovalbumin gene: Evidence for a leader sequence in mRNA and DNA sequences at the exon–intron boundaries, *Proc. Natl. Acad. Sci. U.S.A.* **75**:4853–4857.

Calabretto, B., Robberson, D. L., Barrera-Saldana, H. A., Lambrou, T. P., and Saunders, G. F., 1982, Genome instability in a region of human DNA enriched in *Alu* repeat sequences, *Nature (London)* **296**:219–225.

Calos, M. P., and Miller, J. H., 1980, Transposable elements, *Cell* **20**:579–595.

Cochet, M., Gannon, F., Hen, R., Maroteaux, L., Perrin, F., and Chambon, P., 1979, Organization and sequence studies of the 17-piece chicken conalbumin gene, *Nature (London)* **282**:567–574.

Cooke, N. E., and Baxter, J. D., 1982, Structural analysis of the prolactin gene suggests a separate origin for its 5′ end, *Nature (London)* **297**:603–606.

Cooke, N. E., Coit, D., Weiner, R. I., Baxter, J. D., and Martial, J. A., 1980, Structure of cloned DNA complementary to rat prolactin messenger RNA, *J. Biol. Chem.* **255**:6502–6510.

Cooke, N. E., Coit, D., Shine, J., Baxter, J. D., and Martial, J. A., 1981, Human prolactin: cDNA structural analysis and evolutionary comparisons, *J. Biol. Chem.* **256**:4007–4016.

Corden, J., Wasylyk, A., Buchwalder, P., Sassone-Corsi, D., Kedinger, P., and Chambon, P., 1980, Promoter sequences of eukaryotic protein-coding genes, *Science* **209**:1406–1414.

Daughaday, W. H., 1981, The adenohypophysis, in: *Textbook of Endocrinology* (R. H. Williams, ed.), pp. 73–116, W. B. Saunders, Philadelphia.

DeNoto, F. M., Moore, D. D., and Goodman, H. M., 1981, Human growth hormone DNA sequence and mRNA structure: Possible alternative splicing, *Nucleic Acids Res.* **9**:3719–3730.

Efstratiadis, A., Posakony, J. W., Maniatis, T., Lawn, R. M., O'Connell, C., Spritz, R. A., DeRiel, J. K., Forget, B. G., Weissman, S. M., Slightom, J. L., Blechl, A. E., Smithies, O., Baralle, F. E., Shoulders, C. C., and Proudfoot, N. J., 1980, The structure and evolution of the human β-globin gene family, *Cell* **21**:653–668.

Frohman, L. A., Maran, J. W., and Dhariwal, P. S., 1971, Plasma growth hormone responses to intrapituitary injections of growth hormone releasing factor (GRF) in the rat, *Endocrinology* **88**:1483–1488.

Gubbins, E. J., Maurer, R. A., Lagrimini, M., Erwin, C. R., and Donelson, J. E., 1980, Structure of the rat prolactin gene, *J. Biol. Chem.* **255**:8655–8662.

Hynes, N. E., Kennedy, N., Rahmsdorf, U., and Groner, B., 1981, Hormone-responsive expression of an endogenous proviral gene of mouse mammary tumor virus after molecular cloning and gene transfer into cultured cells, *Proc. Natl. Acad. Sci. U.S.A.* **78**:2038–2042.

Jelinek, W. R., and Schmid, C. W., 1982, Repetitive sequences in eukaryotic DNA and their expression, *Annu. Rev. Biochem.* **51**:813–844.

Klar, A. J. S., Strathern, J. N., Broach, J. R., and Hicks, J. B., 1981, Regulation of transcription in expressed and unexpressed mating type cassettes of yeast, *Nature (London)* **289**:239–244.

Krulich, L., Dhariwal, A. P. S., and McCann, S. M., 1968, Stimulatory and inhibitory effects of purified hypothalamic extracts on growth hormone release from rat pituitary *in vitro*, *Endocrinology* **83**:783–790.

Lee, F., Mulligan, R., Berg, P., and Ringold, G., 1981, Glucocorticoids regulate expression of dihydrofolate reductase cDNA in mouse mammary tumour virus chimaeric plasmids, *Nature (London)* **294**:228–233.

Lerner, M. R., Boyle, J. A., Mount, S. M., Wolin, S. L., and Steitz, J. A., 1980, Are snRNPs involved in splicing?, *Nature (London)* **283**:220–224.

Liebhaber, S. A., Goosens, M. J., and Kan, Y. W., 1980, Cloning and complete nucleotide sequence of human 5′-α-globin gene, *Proc. Natl. Acad. Sci. U.S.A.* **77**:7054–7058.

Martial, J. A., Hallewell, R. A., Baxter, J. D., and Goodman, H. M., 1979, Human growth hormone: Complementary DNA cloning and expression in bacteria, *Science* **205**:602–607.

Martin, T. F. J., and Tashjian, A. H., 1977, Cell culture studies of thyrotropin-releasing hormone action, in: *Biochemical Action of Hormones* (G. Litwack, ed.), pp. 269–312, Academic Press, New York.

Maurer, R. A., 1980, Dopaminergic inhibition of prolactin synthesis and prolactin messenger RNA accumulation in cultured pituitary cells, *J. Biol. Chem.* **255**:8092–8097.

Maurer, R. A., Erwin, C. R., and Donelson, J. E., 1981, Analysis of 5′ flanking sequences and intron–exon boundaries of the rat prolactin gene, *J. Biol. Chem.* **256**:10,524–10,528.

McKenna, M. G., 1969, The origin and early differentiation of therian mammals, *Ann. N. Y. Acad. Sci.* **167**:217–240.

McKnight, S. L., and Kingsbury, R., 1982, Transcriptional control signals of a eukaryotic protein-coding gene, *Science* **217**:316–324.

Moore, D. D., Conkling, M. A., and Goodman, H. M., 1982, Human growth hormone: A multi-gene family, *Cell* **29**:285–286.

Nasmyth, K. A., Tatchell, K., Hall, B. D., Astell, C., and Smith, M., 1981, A position effect in the control of transcription at yeast mating type loci, *Nature (London)* **289**:244–250.

Niall, H. D., Hogan, M. L., Sauer, R., Rosenbaum, I. Y., and Greenwood, F. C., 1971, Sequences of pituitary and placental lactogenic and growth hormones: Evolution from a primordial peptide by gene reduplication, *Proc. Natl. Acad. Sci. U.S.A.* **68**:866–869.

Nicoll, C. S., and Meites, J., 1964, Prolactin secretion *in vitro*: Effects of gonadal and adrenal cortical steroids, *Proc. Soc. Exp. Biol. Med.* **117**:579–581.

Owerbach, D., Rutter, W. J., Martial, J. A., Baxter, J. D., and Shows, T. B., 1980, Genes for growth hormone, chorionic somatomammotropin, and growth hormone-like gene on chromosome 17 in humans, *Science* **207**:289–292.

Owerbach, D., Rutter, W. J., Cooke, N. E., Martial, J. A., and Shows, T. B., 1981, The prolactin gene is located on chromosome 6 in humans, *Science* **212**:815–816.

Page, G. S., Smith, S., and Goodman, H. M., 1981, DNA sequence of the rat growth hormone gene: Location of the 5′ terminus of the growth hormone mRNA and identification of an internal transposon-like element, *Nucleic Acids Res.* **9**:2087–2104.

Robins, D. M., Paek, I., Seeburg, P. H., and Axel, R., 1982, Regulated expression of human growth hormone genes in mouse cells, *Cell* **29:**623–631.

Rogers, J., and Wall, R., 1980, A mechanism for RNA splicing, *Proc. Natl. Acad. Sci. U.S.A.* **77:**1877–1879.

Romero-Herrera, A. E., Lehman, H., Joysey, K. A., and Friday, A. E., 1973, Molecular evolution of myoglobin and the fossil record: A phylogenetic synthesis, *Nature (London)* **246:**389–395.

Rubin, C. M., Houck, C. M., Deininger, P. L., Friedman, T., and Schmid, C. W., 1980, Partial nucleotide sequence of the 300-nucleotide interspersed repeated human DNA sequences, *Nature (London)* **284:**372–374.

Seif, I., Khoury, G., and Dhar, R., 1979, BKV splice sequences based on analysis of preferred donor and acceptor sites, *Nucleic Acids Res.* **6:**3387–3398.

Simpson, E. R., and MacDonald, P. C., 1981, Endocrinology of pregnancy, in: *Textbook of Endocrinology* (R. H. Williams, ed.), pp. 412–422, W. B. Saunders, Philadelphia.

Takahara, J., Arimura, A., and Schally, A. V., 1974, Suppression of prolactin release by a purified porcine PIF preparation and catecholamines infused into a rat hypophysial portal vessel, *Endocrinology* **95:**462–465.

Tashjian, A. H., Yasumura, Y., Levine, L., Sato, G. H., and Parker, M. L., 1968, Establishment of clonal strains of rat pituitary tumor cells that secrete growth hormone, *Endocrinology* **82:**342–352.

Tashjian, A. H., Bancroft, F. C., and Levine, L., 1970, Production of both prolactin and growth hormone by clonal strains of rat pituitary cells: Differential effects of hydrocortisone and tissue extracts, *J. Cell Biol.* **47:**61–70.

Taylor, W. L., Collier, K. J., Weith, H. L., and Dixon, J. E., 1981, The use of aheptadeoxyribonucleotide as a specific primer for prolactin mRNA: A prediction of ambiguous RNA splicing, *Biochem. Biophys. Res. Commun.* **102:**1071–1077.

Tushinski, R. J., Sussman, P. M., Yu, L. -Y., and Bancroft, F. C., 1977, Pregrowth hormone messenger RNA: Glucocorticoid induction and identification in rat pituitary cells, *Proc. Natl. Acad. Sci. U.S.A.* **74:**2357–2361.

Wahli, W., David, I. B., Wyler, T., Weber, R., and Ryffel, G. U., 1980, Comparative analysis of the structural organization of two closely related vitellogenin genes in *X. laevis*, *Cell* **20:**107–117.

Yasumura, Y., Tashjian, A. H., and Sato, G. H., 1966. Establishment of four functional, clonal strains of animal cells in culture, *Science* **154:**1186–1189.

5

Cloning and Structure Analysis of Histocompatibility Class I and Class II Genes

ASHWANI K. SOOD, JULIAN PAN,
PAUL A. BIRO, DENNIS PEREIRA,
VEMURI B. REDDY, HRIDAY K. DAS,
and SHERMAN M. WEISSMAN

1. INTRODUCTION

The major histocompatibility complex in vertebrates consists of a number of closely linked genetic loci that encode a variety of cell-surface glycoproteins and serum proteins known as histocompatibility antigens (Klein, 1975). Through these antigens, the cells of the immune system interact and thus regulate the antibody and cellular immune response to foreign antigens (Zinkernagel and Doherty, 1974; Klein, 1979). Biochemically, these antigens have been subdivided into three classes on the basis of structural and functional homology

ASHWANI K. SOOD, JULIAN PAN, PAUL A. BIRO, DENNIS PEREIRA, VEMURI B. REDDY, HRIDAY K. DAS, and SHERMAN M. WEISSMAN ● Department of Human Genetics, Yale University School of Medicine, New Haven, Connecticut 06510.   Dr. Biro's present address is Biological Laboratories, Harvard University, Cambridge, Massachusetts 02138.   Dr. Pereira's present address is Microbiological Genetics, Pfizer Central Research, Groton, Connecticut 06340.   Dr. Reddy's present address is Integrated Genetics, Framingham, Massachusetts 01701.

(Ploegh et al., 1981). Thus, the class I antigens (HLA-A, B, C; H2K, D, L) are cell-surface glycoproteins composed of two polypeptide subunits. The heavy chain carries the antigenic determinants and is in noncovalent association with β_2-microglobulin. Similarly, the class II antigens are also composed of two polypeptide chains. Significant features of these antigens are that they are highly polymorphic within the population. At least 35 alleles are known at the *HLA-B* locus, 20 at the *HLA-A* locus, and fewer than 10 at the *HLA-C* locus. In the mouse, the extent of polymorphism at the class I loci is even higher; i.e., close to 100 alleles at both the *H2-K* and *-D* loci. In contrast, the polymorphism at the *HLA-D* locus is only beginning to be uncovered with recent definitions of multiple loci encoding these antigens (Tanigachi et al., 1980; Shaw et al., 1981; Shackelford et al., 1981).

A further interesting feature of these antigens concerns their tissue distribution. While the class I antigens are present on all tissues except sperm, the class II antigens show a very limited distribution in that they are present on a subset of the cells of lymphoid tissue (Williams et al., 1980). Finally, the studies on the expression of the class I antigens on the surface of differentiating teratocarcinoma cell lines have shown that the expression of class I antigens is developmentally regulated (Miller and Ruddle, 1977; Gmur et al., 1980; Goodfellow et al., 1982).

We became interested in cloning the genes encoding class I and class II antigens in order to study the underlying mechanisms responsible for the polymorphism exhibited by these antigens. A further interest in class I histocompatibility genes is to study the molecular mechanisms that regulate their developmental expression as well as differential levels of expression in different tissues.

2. CLONING THE CLASS I AND CLASS II HISTOCOMPATIBILITY GENES

Our approach to cloning the histocompatibility genes envisaged the use of synthetic oligonucleotides that were predicted to be complementary to a sequence in the histocompatibility messenger RNA (mRNA). Such oligonucleotides serve as specific primers in a primer-dependent complementary DNA (cDNA) synthesis reaction. The cDNA products were then separated on the acrylamide–urea gels and individual cDNA products characterized to identify *HLA*-specific cDNA product. Finally, the *HLA*-specific cDNA product was used as a probe to screen the cDNA and genomic libraries.

B7, an oligodeoxynucleotide was synthesized. Our attempts to characterize the primer-specific cDNA products obtained from the full-length cDNA synthesis reaction were unsuccessful, due to the poor resolution of the large-sized cDNA product on the acrylamide–urea gel (Sood *et al.,* 1981). To overcome this problem, we developed a highly efficient method that reduces the size and heterogeneity of the cDNA products. In short, the cDNA synthesis was performed with the labeled primer on total polyadenylic acid [poly(A)] mRNA in the presence of three deoxynucleoside triphosphates and a dideoxynucleoside triphosphate. The resulting cDNA products were separated according to the chain length on acrylamide–urea gels. The principle of the method and its use in cloning the cDNA and specifying a class I HLA-B antigen has been published (Sood *et al.,* 1981). We describe below the use of this approach to isolate the gene encoding the heavy chain of class II antigens.

Figure 1 shows the amino acid sequence of the heavy chain of a class II antigen and the predicted nucleotide sequences at a unique site in the mRNA. Earlier experience suggested that the mismatch near the 5′ end of the primer did not appreciably decrease the cDNA synthesis from class I histocompatibility mRNA (unpublished results). We therefore synthesized the four oligodeoxynucleotide primers as shown in Fig. 1. From examination of the pre-

A. HLA-DR p34 protein: NH₂ Val ll   e ll   e Gl   n Al   a <u>Glu Phe Tyr Leu</u>————

B.   p34 mRNA   5′ GUNAUU AUU CA GGC   U GA G UU U UA U CU————
                                   C   C   A   C   A   C   C
                                   A   A   A
                                                     G

C.   p34 cDNA   3′   AGTCCGG <u>CTC AAG ATG GA</u>⁵′

                              D.   Primers synthesized:
                                        ³′CT<u>T</u>AA<u>A</u>ATGGA⁵′ Primer 1
                                        CT<u>C</u>AA<u>A</u>ATGGA
                                        CT<u>T</u>AA<u>G</u>ATGGA
                                        CT<u>C</u>AA<u>G</u>ATGGA Primer 4

FIGURE 1.   (A) Limited amino acid sequence of p34 polypeptide; (B) predicted mRNA sequence of p34 mRNA. (C) Nucleotide sequence of the 18-nucleotide-long extension product (the primer part is underlined). (D) The four oligonucleotides complementary to the predicted sequence in the *Alu* mRNA.

dicted nucleotide sequence in the mRNA upstream from the site of primer hybridization, it was apparent that a U residue can occur only at 7 or 8 residues upstream of the 3' end of the primer. Consequently, the cDNA synthesis was performed in the presence of dideoxyadenosine triphosphate and the three other deoxynucleoside triphosphates. Figure 2 shows the autoradiograph of the cDNA synthesis reaction using primers 1 and 4. It is evident that the 18- and 19-residue-long bands are present only in the extension performed using primer 4. Therefore, subsequent cDNA synthesis reaction was performed using only primer 4. The 18- and 19-residue-long bands were characterized by the two-dimensional electrophoresis–homochromatography techniques (data not shown). The 18-nucleotide-long band showed the nucleotide sequence complementary to the predicted nucleotide sequence of class II mRNA. From the knowledge of this additional sequence, a 20-nucleotide-long oligonucleotide was synthesized. The 20-mer was expected to be a highly specific primer. The result of primer extension using the 20-mer primer is shown in Fig. 3. The most intense band, about 170 bases long, contained the nucleotide sequence that matched with the rest of the N-terminal amino acid sequence of the class II antigen heavy chain (Fig. 4). The 170-base-long cDNA was used as a probe to screen a genomic λ library made from the DNA obtained from the JY cell line. The genomic clones containing the complete gene were isolated and characterized (Das *et al.*, 1984).

From these results and those published earlier, it appears that specific oligodeoxynucleotide-primed, didioxynucleoside-triphosphate-terminated limited cDNA synthesis is a highly sensitive method to obtain cDNA probes to clone mammalian genes that are expressed at low levels. Some of the advantages of this method are as follows:

1. Very limited amino acid sequence is required. A stretch of ten amino acids is usually sufficient both to derive the nucleotide sequence of the primer and to predict the length of the desired cDNA product in an appropriate dideoxynucleoside-triphosphate-terminated cDNA synthesis reaction. With the advancement of the technology for peptide sequencing, such information should be available for scarce proteins. Furthermore, the protein sample need not be pure as long as the desired protein is the major constituent of the mixture. This is in contrast to other procedures that utilize *in vitro* translation followed by antibody precipitation, in which pure protein samples might be required to produce specific antisera.

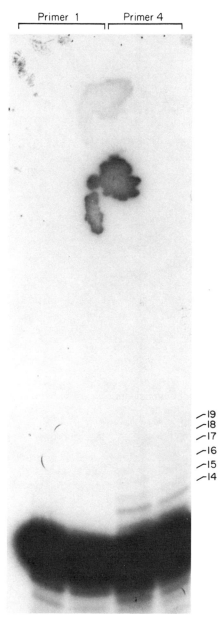

FIGURE 2. Autoradiograph of the cDNA synthesis reaction performed with primer 1 (two left lanes) and primer 4 (two right lanes) in the presence of dideoxyadenosine triphosphate and three other deoxynucleoside triphosphates. The numbers alongside the bands indicate the chain length of the respective bands.

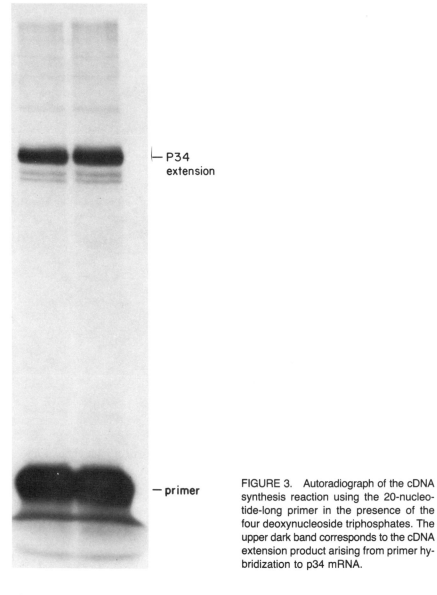

FIGURE 3.   Autoradiograph of the cDNA synthesis reaction using the 20-nucleo-tide-long primer in the presence of the four deoxynucleoside triphosphates. The upper dark band corresponds to the cDNA extension product arising from primer hybridization to p34 mRNA.

```
5' ppp NNN~29 GACTCCCAACAGAGCGCCCAAGAAGAAAATGGCCATAAGTGGAGTC
                                              Met Ala Ile Ser Gly Val

    CCTGTGCTAGGATTTTTCATCATAGCTGTGCTGATGAGCGCTCAGGAATCA
    Pro Val Leu Gly Phe Phe Ile Ile Ala Val Leu Met Ser Ala Gln Glu Ser
    ――――――――――――――― Leader Peptide ―――――――――――――――

    TGGGCTATCAAAGAAGAACATGTGATCATTCAGGCC GAGTTCTACCT  3'
    Trp Ala Ile Lys Glu Glu His Val Ile Ile Gln Ala Glu Phe Tyr Leu
```

FIGURE 4.   Nucleotide sequence of the p34 cDNA product and its identity. The first 20 N-terminal amino acid residues of the p34 are shown. The predicted amino acid sequence of the signal peptide is underlined.

2. As a result of very limited cDNA synthesis, opportunities for pre-mature termination of reverse transcriptase are avoided. The method therefore yields the maximum synthesis of cDNA products.* This property of the method makes it especially suitable for cloning the genes that encode low abundant messages.

3. The cDNA products are less heterogeneous.

4. The size of the desired cDNA product can be predicted. This generally reduces the number of cDNA products to be characterized to less than 10%.

3. STRUCTURE OF THE CLASS I *HLA-B7* COMPLEMENTARY DNA CLONE

The complete nucleotide sequence of the *HLA-B7* cDNA clone was determined by the chemical degradation method of Maxam and Gilbert (1977). The cDNA clone specifies 300 of the 338 amino acids of the heavy chain followed by 425 base pairs (bp) of the 3′ untranslated sequence, including the AAUAAA-poly(A) addition signal (Proudfoot and Brownlee, 1976). The comparison of the nucleotide sequence of this cDNA clone with the nucleotide sequence of an *HLA* gene (LN11A), in addition to revealing the intron–exon organization of the gene, showed that the 3′ untranslated region of HLA mRNA is encoded on a separate exon. Similar comparison with another *HLA* gene confirmed the foregoing observations. The significance of this obser-vation to the possibility of polymorphism at the carboxy terminus of HLA antigens will be discussed later.

Comparison of the translated nucleotide sequence of the coding part of the cDNA clone with the amino acid sequence of the heavy chain of HLA-B7 antigen is shown in Fig. 5. The two differ at 2 of 300 amino acids; one of the differences could be an artifact of protein sequencing. Another differ-ence at amino acid 273, from Arg to Gly, is situated a few residues before

---

* We have performed the cDNA synthesis reaction in the presence of only three deoxynucleoside triphosphates and in the absence of dideoxynucleoside triphosphate. Such cDNA synthesis reactions showed extra bands in addition to those present in the normal dideoxynucleoside-triphosphate-terminated reaction (our unpublished results). One explanation is that the additional cDNA products result from the reextension of some short cDNA products that, after extension from one mRNA species, dissociated and reextended from another RNA species. This suggests that an individual mRNA species may be utilized as a template more than once, thus leading to increased cDNA synthesis. Such an opportunity is present only in the limited cDNA synthesis reaction.

```
                                    40                    50                    60                    70
B7 cDNA         GCCGCGAGTCCGAGGAGAGGAGGCCGCGGCGCCGTGGATAGAGCAGGAGGGGCCGGAGTATTGGGACCGGAACACAGATCTACAAGGCCCAAGCACAGACTGACCGAGAG
B7 Heavy Chain   A  A  S  P  R  E  E  P  R  A  P  W  I  E  Q  E  G  P  E  Y  W  D  R  N  T  I  Y  K  A  Q  A  Q  T  D  R  E

                       80                    90                   100                   110
                AGCCTCGGAACCTGCGCGGCTACTACAACGAGACGAGGCTGGGTCTCACACCCTTCACACCTTCAGAGCATGTACGGTTGCCACGTGGGCGCCGACGGGCGCCTCCTCCGCGGGCAT
                 S  L  R  N  L  R  G  Y  Y  N  Q  S  E  A  G  S  H  T  L  Q  S  M  Y  G  C  D  V  G  P  D  G  R  L  L  R  G  H

                       120                   130                   140                   150
                GACCAGTACGCCTACGACGGCAAGGATTACATCGCCCTGAACGAGGACCTGCGCTCCTGGACCGCCGCGGACACCGCTGCGCAGATCACCCAGCGCAAGTGGGAGGCGGCC
                 D  Q  Y  A  Y  D  G  K  D  Y  I  A  L  N  E  D  L  R  S  W  T  A  A  D  T  A  A  Q  I  T  Q  R  K  W  E  A  A

                       160                   170                   180
                CGTGAGGCGGAGCAGCGGAGAGCCTACCTGGAGGGCGAGTGCGTGGAGTGGCTCCGAAGATACCTGGAGAACGGAAGGACAAGCTGGAGCGCGCTGACCCCTAAGACA
                 R  E  A  E  Q  R  R  A  Y  L  E  G  E  C  V  E  W  L  R  R  Y  L  E  N  G  K  D  K  L  E  R  A  D  P  P  K  T

                       190                   200                   210                   220
                CACGTGACCCACCACCCCATCTCTGACCATGAGGCCACCCTGAGGTGCTGGGCCCTTGGTTTCTATCCTGCGGAGATCACACTGACCTGGCAGCGGGATGGCGAGGACCAA
                 H  V  T  H  H  P  I  S  D  H  E  A  T  L  R  C  W  A  L  G  F  Y  P  A  E  I  T  L  T  W  Q  R  D  G  E  D  Q

                       230                   240                   250                   260
                ACTCAGGACACTGAGCTTGTGGAGACCAGACCAGCAGGAGATAGAACCTTCCAGAGTGGGGCAGCTGTGGTGCCATCTGAGAAGAGAGCAGAGATACACATGCCATGTA
                 T  Q  D  T  E  L  V  E  T  R  P  A  G  D  R  T  F  Q  K  W  A  A  V  V  V  P  S  G  E  E  Q  R  Y  T  C  H  V
                                                        E

                       270                   280                   290
                CAGCATGAGGGGCTGCCAAGCCACTNACACTGGGATGGGAGCGTCTTCCCAGTCAACGTTGCTGGCATTGTTGCTGGCCGGCTGTCCTAGCAGTTGTGGTC
                 Q  H  E  G  L  P  K  P  L  T  L  G  W  E  P  S  S  Q  S  T  V  P  I  V  A  G  P  A  V  L  A  V  V  V
                                       R                                          *              *

                       300                   310                   320                   330
                ATCGGAGCTGTGGTCGCTGCTGTGATGTGTAGGAGGAAGAGTTCAGGTGGAAAAGGAGGAGGGTCAGGGGAAGCCGGCAGAGTGCCGAGCAGTGCCAGGGCTCTGATGTGTCT
                 I  G  A  V  V  A  A  V  M  C  R  R  K  S  S  G  G  K  G  G  S  Y  S  Q  A  A  C  S  D  S  A  Q  G  S  D  V  S
                 *

                       338
                CTCACAGCTTGA
                 L  T  A  TER.
```

FIGURE 5. *Top line:* Nucleotide sequence of the HLA-B7 cDNAs. *Middle line:* Translated sequence of HLA-B7 cDNA. *Bottom line:* Positions at which the amino acid sequence of HLA-B7 heavy chain differs from the translated amino acid sequence of HLA-B7, e.g., positions 242 and 273. (*) Positions in the heavy-chain amino acid sequence that were unassigned.

its transmembrane hydrophobic domain. This difference is unlikely to contribute to the change in the antigenic determinants of the molecule. This comparison clearly establishes that the cDNA clones specify the heavy chain of an HLA-B7 antigen. Furthermore, from the nucleotide sequence of this clone, amino acid residues have been assigned at five positions in the amino acid sequence of HLA-B7 protein (Ploegh et al., 1981). These residues are Ile, Ile, Pro, Leu, and Ile at positions 284, 287, 291, 294, and 299, respectively.

4. COMPARISON OF THE NUCLEOTIDE SEQUENCE OF THE *HLA-B7* COMPLEMENTARY CLONE WITH THE NUCLEOTIDE SEQUENCE OF THE CODING REGIONS OF AN *H-2* GENE

On the basis of comparison of the amino acid sequence of HLA antigens with H-2 antigens, it was observed that while the differences between the two molecules are scattered in the three extracellular domains, the two molecules are highly homologous (>75%) in those domains. In contrast, the transmembrane and cytoplasmic domains in the two molecules appeared highly divergent (Ploegh et al., 1981; Coligan et al., 1981). At the nucleotide-sequence level (Fig. 6), however, it is clear that the two molecules are homologous not only in the exons specifying the three extracellular domains, but also in the exons specifying the transmembrane and the cytoplasmic domains, except the three insertions/deletions of 3, 3, and 9 bp, respectively (see Fig. 6), that do not lead to a change in the reading frame. This supports the argument that the cytoplasmic domains have a critical role in the overall function of histocompatibility antigens (Pober and Strominger, 1981; Pober et al., 1981).

Further comparison of several *HLA* and *H-2* genes showed an identical deletion/insertion pattern in the transmembrane and cytoplasmic exons (unpublished results). These results strongly support the model that the histocompatibility gene duplication or conversion occurred subsequent to speciation. The alternative possibility that gene duplication preceded speciation cannot be ruled out because of gene homogenization mechanisms giving rise to the "human type" and "mouse type" sequences (see below and Evans et al., 1982; Lopez de Castro et al., 1982). A further outcome of the comparison of the nucleotide sequence is that although the HLA cDNA and genes show

EXON 2

```
HLA B7 cDNA   GCC GCG AGT AGA GAG GAG CCG GCG CCG TGG ATA GAG CAG GAG GGG CCG GAG TAT TGG GAC CGG AAC ACA CAG ATC
H2L Gene      **G *A* *A* T*T *** A** *** A** **G *** *** *** *** *** **G *** *** **G *** *T* **G *** ***

              TAC AAG GCC CAA GCA CAG ACT GAC CGA GAG AGC CTG CGC CGC TAC TAC AAC CAG AGC GAG GCT G
              GC* *** *G* **G *AG *** TGG TT* *** *T* *A* *** *T* *T* *** *** *** **C* *C* *GC* *
```

EXON 3

```
              GG TCT CAC ACC CTT CAG AGC ATG TAC GGT TGC GAC GTG GGG CCG GAC CTC CTC CGC GGG CAT GAC CAG TAC GCC
              *C A** *** **A **C *** T*G *** **C **T *** T** **T *** P** *** *** *** TAC **G *** *TC ***

              TAC GAC GGC AAG GAT TAC ATC GCC CTG AAC GAG GAC CAG CGG ACC TGG TGG ACC GAC GAC ACG GGC GAG ATC ACC ACC CAG
              *** *** *** TGC *** *** *** *** *** **A *** **A A*G **G TCC A*G TT* **G *T* T*G **G *** *** ***  *GA

              CGC AAG TGG GAG GCG GCC CGT GAG GCG GAG CAG CGG GAG GGC GAG TGC CTC CGA AGA TAC
              *** *** *** CA* **T G** *CT **A T*T TAC **G *** *** *** *** **A *AC *** ***

              CTG GAG AAC GGG AAG GAC AAG CTG GAG CGC GCT G
              *** A** *** **T *CT *C* *** CT* *** A*A *
```

EXON 4

```
              AC CCC CCT AAG ACA CAC GTG ACC CAC CCC ATC TCT GAC CAT GAG GCC ACC CTG AGG TGC TGG GCC CTT GGT TTC TAT
              *T T** **A G** **T *** **T *** *** **T *** *GA *** A*A GG* **A *T* *** *** **G **C *** **G ***  **C

              CCT GCG GAG ATC ACA CTG ACC TGG CAG CGG GAT GGC GAG GAC CAA ACT CAG GAC CTT GTG GAG ACC AGA CCA GCA
              *** **T **C *** *** *** *** TT* A** **G *** **G *TG *** *** *TG *** *** *** **G **T ***

              GGA GAT AGA ACC TTC CAG AAG TGG GCA GCT GTG GTG TGT CCA TCT GGA GAA GAG CAG AGA TAC ACA TGC CAT GTA CAG CAT
              **G G** *** *** *** *** *** T** *** T*T *** *** **T C** **G A*G A*G *** *AT *** *** *G* **G T*C ***

              GAG GGG CTG CCG AAG CCA CTN ACA CTG GGA TGG G
              *** *** **T G** **C **G **C A** **** *
```

EXON 5

```
AG CCG TCT TCC CAG TCC ACC GTC CCC ATC GTT GGC ATT GTT GCT GGC CCG GCT GTC CTA GCA GTT GTG GTC ATC   GGA
** **T **C C*G ===    *** **T *A* T*T TA* A*G *TG **C *** *TT *T* *G* *** **T A** *C* *** ATT***
                                                                                           ===

GCT GTG GTC GCT GTG ATG TGT AGG AGG AAG AGT TCA G
*** *** **G *** TT* *** *** AAG **A *** *GA *AC A** *
```

EXON 6

```
GT GGA AAA GGA GGG AGC TAC TCT CAA GCT GCG T
** *** *** *** *** GA* **T G** *TG *** C*A G
```

EXON 7

```
GC AGC GAC AGT GCC CAG GGC TCT GAT GTG TCT CTC ACA GCT TGA AAA G
**         T** *** A** *** **A A** *** *** CG* *A* **T*** *
   =========
```

EXON 8

```
-----------3'  untranslated-------
CG TGA ------3'- untranslated-------
```

FIGURE 6. Comparison of the nucleotide sequence of HLA-B7 cDNA with the coding sequence of *H2L*^{d} gene. (*) Identity between the two sequences. In Exon 5, there are two 3-bp deletions/insertions double-underlined. Similarly, in Exon 7, there is a 9-bp deletion/insertion double-underlined.

a 9-bp insertion in the beginning of the second cytoplasmic exon, a mutation in the *HLA* gene has led to a termination codon 9 bp upstream of the termination codon that is utilized in the *H-2* genes. The consequence of this difference is that while the 3′-terminal exon of the *H-2* gene specifies the two carboxy-terminal amino acids and the termination codon (Steinmetz *et al.*, 1981; Moore *et al.*, 1982; Evans *et al.*, 1982), the 3′-terminal exon of the *HLA* genes does not contribute to the structure of HLA peptide. One possibility is that the function of the 3′-terminal exon in the *H-2* gene was simply to juxtapose a termination codon in the reading frame and with the appearance of a termination codon in the subterminal exon in the *HLA* genes, the 3′-terminal exon became unnecessary. This observation is consistent with the lack of the donor splice site downstream from the end of the subterminal exon in one of the *HLA* genes (Malissen *et al.*, 1982). On the other hand, the presence of the aforementioned donor splice site in three of the *HLA* genes (Fig. 5 and unpublished results), although not inconsistent with the preceding argument, does raise the possibility that the terminal exons of *HLA* genes might specify an additional carboxy-terminal domain of an altered HLA antigen. Such a polymorphism at the carboxy terminus of the histocompatibility antigens might be species-specific, since the 3′-terminal exons in *HLA* and *H-2* genes are more divergent than the remaining exons.

5. STRUCTURE OF A CLASS I HISTOCOMPATIBILITY PSEUDOGENE

Using the HLA-B7 cDNA clone as a probe, we screened a genomic DNA library constructed from a human placenta of unknown HLA type. A strongly hybridizing clone (LN11) was isolated, and a part of it was subcloned and designated *LN11A* (P. A. Biro *et al.*, unpublished results). The nucleotide sequence of a stretch of 6 kb of this subclone was determined by the Maxam–Gilbert procedure and compared with the nucleotide sequence of HLA-B7 cDNA clone as shown in Fig. 7 and the *HLA-B7* genomic clone (unpublished results). Figure 7 also shows the translated amino acid sequence encoded by the gene. From this comparison, and unpublished results, we find that the gene is split into eight exons. Exon 1 encodes the 5′ untranslated region of HLA mRNA, together with the signal peptide; Exons 2, 3, and 4 specify the three extracellular domains of the mature heavy chain. Exon 5 encodes the transmembrane hydrophobic domain, followed by Exons 6 and

7, which together encode the cytoplasmic domain. Exon 8 encodes the 3′ untranslated region of HLA mRNA. The nucleotide sequence of *LN11A* between Exons 5 and 6 showed the presence of potential donor and acceptor splice sites that, if utilized, would give rise to polymorphism at the carboxy terminus of HLA antigens (discussed later).

Comparison of the nucleotide sequence of Fig. 6 revealed that *LN11A* contains several deletions and insertions in most of the exons. These mutations give rise to large shifts in the reading frame, as well as result in several termination codons. These results show that *LN11A* is a pseudogene. A closer look at the locations of mutations in the *LN11A* gene has revealed an interesting pattern, in that the majority of the deletions in *LN11A* occur in the first three exons, while the next four exons contain only one deletion and three insertions.

Comparison of the nucleotide sequence of *LN11A* with that of another *HLA* gene (Malissen *et al.*, 1982) summarized in Fig. 8 (unpublished results) confirmed the observations cited above.

As shown in Fig. 8, in Exon 1, *LN11A* has two deletions of 1 and 2 bp, respectively. In addition, a termination codon appears in the reading frame.

In Exon 2, *LN11A* has three deletions of 3, 1, and 1 bp, respectively; the latter two deletions result in large shifts in reading frame.

In Exon 3, again, *LN11A* has five deletions of 5, 2, 1, 1, and 1 bp, respectively, that again result in two termination codons in addition to the large shifts in the reading frame.

In Exon 4, *LN11A* has a 1-bp deletion followed by a 1-bp insertion only 4 bp after the deletion. The net effect is a change of two amino acids.

In Exon 5, *LN11A* has two 1-bp insertions. The first insertion changes the nature of the peptide domain encoded by this exon. Sixty nucleotides after the first insertion, a second insertion of 1 bp further changes the reading frame that results in a termination codon 7 bp after this insertion. Due to the first frameshift, this exon now specifies a relatively hydrophilic segment of 24 amino acids.

Exons 6 and 7 are identical in size in the two genes. The exon that specifies a 3′ untranslated region of HLA mRNA is about 90% homologous for the first 300 nucleotides. This is followed by a deletion of 80 bp in *LN11A*, to be followed by a homologous segment of about 50 bp that includes the polyadenylation. From these comparisons, it is clear that Exons 1, 2, and 3 contain several shifts in the reading frame that are due to deletions in *LN11A*. In addition, three termination codons appear in these exons.

In contrast, Exon 4 appears to have only two amino acid substitutions,

LN11A gene

EXON 1

EXON 2

EXON 3

EXON 4

B7 CDNA

150
310
470
630
790
950
1106
1267
1417
1557
1721
1868
2011
2163
2327
2492
2649
2805
2949

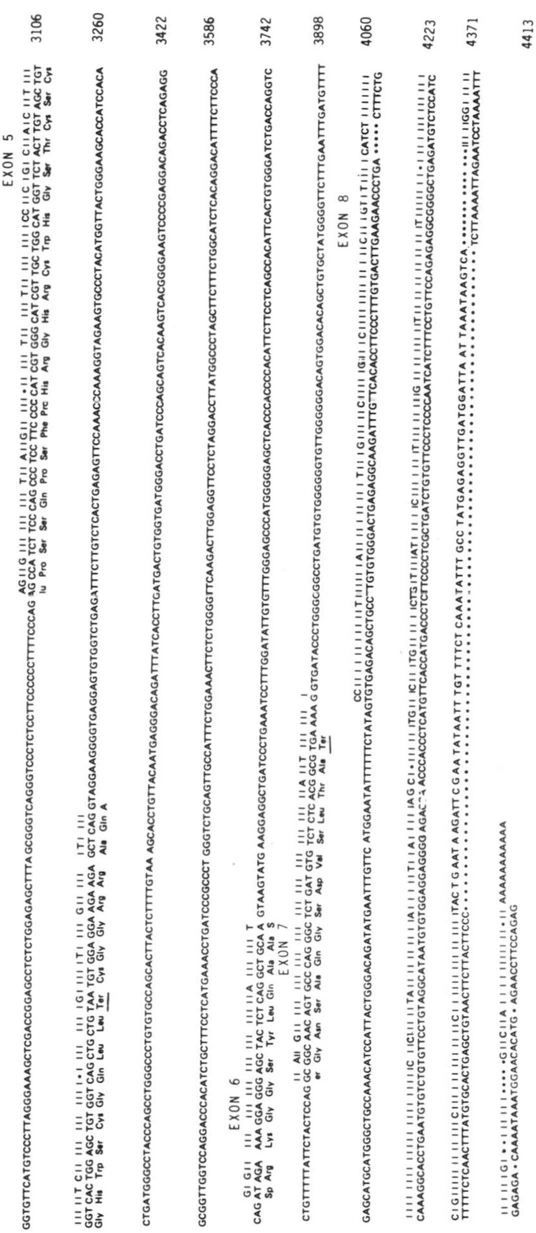

EXON 5

EXON 6

EXON 7

EXON 8

FIGURE 7. Comparison of the nucleotide sequence of HLA-B7 cDNA with that of *LN11A* histocompatibility gene. The translated nucleotide sequence of the gene is also shown. (*) Deletion in the appropriate sequence. Termination codons are underlined.

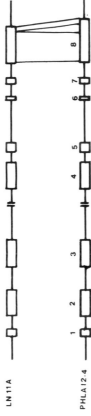
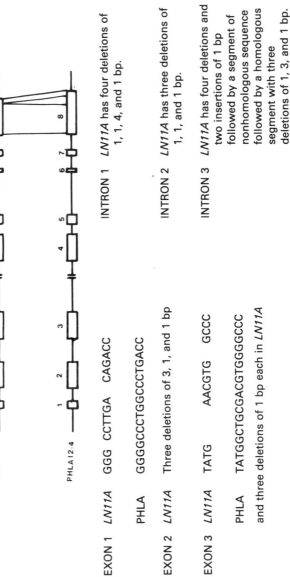

EXON 1   *LN11A*   GGG CCTTGA   CAGACC

           PHLA   GGGGCCCTGGCCCTGACC

EXON 2   *LN11A*   Three deletions of 3, 1, and 1 bp

EXON 3   *LN11A*   TATG   AACGTG   GCCC

           PHLA   TATGGCTGCGACGTGGGGCCC

           and three deletions of 1 bp each in *LN11A*

INTRON 1   *LN11A* has four deletions of 1, 1, 4, and 1 bp.

INTRON 2   *LN11A* has three deletions of 1, 1, and 1 bp.

INTRON 3   *LN11A* has four deletions and two insertions of 1 bp followed by a segment of nonhomologous sequence followed by a homologous segment with three deletions of 1, 3, and 1 bp.

EXON 4 *LN11A* has a single-bp deletion followed by a 1-bp insertion 4 bp after the deletion.

EXON 5 *LN11A* has two 1-bp insertions. The first insertion leads to a hydrophyllic segment of 24 amino acids.

EXON 6 Identical

EXON 7 Identical

EXON 8 The first 300 bp are homologous. Then *LN11A* has a deletion of 70 bp flanked by TCTTAAAATT followed by a 500-bp homologous segment.

INTRON 4 *LN11A* has a deletion of 1 bp and three insertions of 1, 3, and 1 bp.

INTRON 5 *LN11A* has four 1-bp deletions and one 2-bp insertion.

INTRON 6 *LN11A* has a 1-bp deletion and a 1-bp insertion.

INTRON 7 *LN11A* has two deletions of 1 and 2 bp.

FIGURE 8. Summary of comparison of the nucleotide sequence of *LN11A* with that of pHLA 12'4.

due to a deletion/insertion. This exon of *LN11A* therefore most likely specifies a functional domain. Exon 5 has a frameshift mutation due to two 1-bp insertions. The first insertion changes the reading frame such that the following 20 residues are relatively rich in hydrophilic amino acids, in contrast to the usual hydrophobic amino acids present in the transmembrane domain. The second insertion aligns a termination codon in this altered reading frame. The net result of these two insertions is that Exon 5 in *LN11A* specifies for a hydrophilic domain. A deletion in this part of an H-2 cDNA clone leading to a frameshift mutation that is followed by a termination codon was observed by others, and it was suggested that such a hydrophilic domain might be part of a secreted H-2 antigen (Cosman *et al.*, 1981).

In view of this reasoning, it is assumed that Exon 4 specifies a functional domain in *LN11A*. Finally, the next two exons that encode the two cytoplasmic domains are normal in that they lack mutations. These observations point to a curious asymmetry in the nature and location of mutations in *LN11A*. In short, it appears that the exons encoding the signal peptide and the first two extracellular domains are mutated, while Exons 4, 5, 6, and 7 appear normal.

Second, the nature of mutation in Exons 1, 2, and 3 is due solely to deletions in *LN11A*. In contrast, in Exons 4, 5, 6, and 7, the nature of mutation is predominantly due to insertions (75%) in *LN11A*. From the comparison of the amino acid sequence between *HLA-A2* and *HLA-B7* (Orr *et al.*, 1979) and from the biochemical localization of mutations in the H2K^b allele (Nairn *et al.*, 1980), it appeared that the two extracellular domains specified by Exons 2 and 3 exhibit antigenic determinants of the molecules, but recent results suggest that the N-terminal portion of the third extracellular domain specified by Exon 4 might specify a relatively exposed antigenic determinant (Lopez de Castro *et al.*, 1982). In contrast, Exons 4, 5, 6, and 7 specify the relatively constant framework domains of the molecule. Furthermore, from the comparison of the introns between the two genes, it can be seen that apart from a few deletions/insertions, the introns are of similar size and show about 80% homology throughout, except within a short region in the intron separating Exons 3 and 4.

In summary, therefore, we find that Exons 1, 2, and 3 are severely mutated, while Exons 4, 5, 6, and 7 are normal, and a short segment of nonhomologous sequence separates the former from the latter. From these results, we hypothesize that Exons 4, 5, 6, and 7 in *LN11A* have been corrected by a process such as gene conversion (Baltimore, 1981) and that one of the boundaries of this gene-conversion process might fall within the nonhomologous sequence. Such a gene-conversion event may be important in preventing the decay of the exons that predominantly specify the framework domains, while allowing the mutations to accumulate in the exons that specify the domains that predominantly carry the polymorphic antigenic determinants. The evidence cited above for the gene conversion among histocompatibility genes is qualitatively different from that presented by others (Evans *et al.*, 1982; Lopez de Castro *et al.*, 1982).

6. POLYMORPHISM AT THE CARBOXY TERMINUS OF HISTOCOMPATIBILITY ANTIGENS

As mentioned above, the nucleotide sequence of *LN11A* revealed the presence of potential donor and acceptor splice sites between Exons 5 and 6. Four of the five genes, including gene *27.5,* in which the potential donor has an unassigned base, contain a potential donor site 10 or 11 bp downstream from the known donor site (Fig. 9). Similarly, all the five genes contain a potential acceptor site 15–20 bp upstream of the known acceptor. In the case of the *HLA-B7* gene, a third potential acceptor site exists. These potential splice sites, if utilized *in vivo,* would alter the reading frames of the cytoplasmic exons and hence give rise to polymorphic structure at the carboxy terminus of HLA antigens. In gene *27.1,* however, the potential donor site occurs upstream from the normal donor site. It is noteworthy that this potential donor site includes the first internal termination codon of this gene, which characterized it as a pseudogene or a gene that encodes an HLA antigen lacking the cytoplasmic domains (Steinmetz *et al.*, 1981). If, on the other hand, the upstream potential donor site is utilized in this gene, it would then encode an HLA antigen that has a cytoplasmic domain. Expression studies utilizing this cloned gene might be illuminating. In this regard, it seems important to determine the structure of the gene product(s) by studying the expression of individual cloned histocompatibility genes in different cell types.

7. SUMMARY

We have described a highly sensitive method in which synthetic oligonucleotides were used to clone both the class I and class II histocompatibility genes. This method should be generally applicable for cloning mammalian

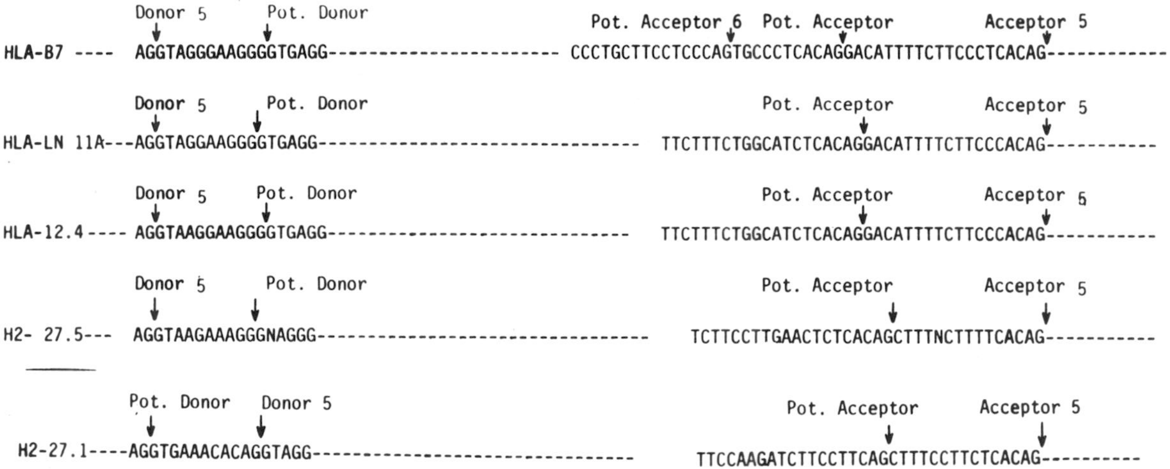

FIGURE 9. Comparison of the nucleotide sequence in the vicinity of the donor and acceptor splice sites located between Exons 5 and 6. (↓) Normal and potential splice sites.

genes that are expressed at low levels. The nucleotide sequence of the class I cDNA clone is identical to the amino acid sequence of *HLA-B7* heavy chain except at two residues. Comparison with the *H-2* coding sequence points to a divergent evolution of the histocompatibility genes in the two species. The structure of a class I histocompability gene and its comparison with the cDNA clone showed that the gene is split into eight exons, which largely correspond to the protein domains predicted from the amino acid sequence except that the cytoplasmic domain is specified by two exons followed by another exon that specifies the 3' untranslated region of the HLA mRNA. The nucleotide sequence of this gene also revealed that it is a pseudogene, and its comparison with another *HLA* gene suggests the existence of a gene-conversion mechanism by which parts of a gene are converted by another.

The comparison of the nucleotide sequence of several *HLA* and *H-2* genes showed that potential splice site sequences occur in the vicinity of the donor and acceptor splice sites between the transmembrane and the first cytoplasmic exons. This observation has led us to the hypothesis that the carboxy terminus of HLA antigens might also be polymorphic. Critical evidence is needed before one might speculate on the mechanistic basis of such a polymorphism.

REFERENCES

Baltimore, D., 1981, Gene conversion: Some implications for immunoglobin genes, *Cell* **24**:592–594.

Coligan, J. E., Kindt, J. J., Uehara, H., Martinko, J., and Nathenson, S. G., 1981, Primary structure of a murine transplantation antigen, *Nature (London)* **291**:35–39.

Cosman, D., Khoury, G., and Jay, G., 1981, Three classes of mouse H-2 messenger RNA distinguished by analaysis of cDNA clones, *Nature (London)* **295**:73–76.

Das, H. K., Lawrance, S., and Weissman, S. M., 1984, Nucleotide sequence of the gene encoding the α-subunit of HLA-DR polypeptide, *Proc. Natl. Acad. Sci. U.S.A.* (submitted).

Evans, G. A., Margulies, D. H., Camerini-Otero, R. D., Ozato, K., and Seidman, J. G., 1982, Structure and expression of a mouse major histocompatibility antigen gene H2L^d, *Proc. Natl. Acad. Sci. U.S.A.* **79**:1994.

Gmur, G., Solter, D., and Knowles, B. B., 1980, Independent regulation of H2K and HsD gene expression in murine teratocarcinoma cell hybrids, *J. Exp. Med.* **151**:1349–1359.

Goodfellow, P. N., Banting, G., Trowsdale, J., Chambers, S., and Solomon, E., 1982, Introduction of a human X-6 translocation chromosome into a mouse teratocarcinoma: Investigation of control of HLA-A, B, C expression, *Proc. Natl. Acad. Sci. U.S.A.* **79**:1190–1194.

Klein, J., 1975, *Biology of the Mouse Histocomptability Complex*, Springer-Verlag, Berlin.

Klein, J., 1979, The major histocompatibility complex of mouse, *Science* **203**:516–521.

Lopez de Castro, J. A., Strominger, J. L., Strong, D. M., and Orr, H. T., 1982, Structure of cross-reactive human histocompatibility antigens HLA-A28 and HLA-A2: Possible implications for the generation of HLA polymorphism, *Proc. Natl. Acad. Sci. U.S.A.* **79**:3813–3817.

Malissen, M., Malissen, B., and Jordan, B. R., 1982, Exon/intron organization and complete nucleotide sequence of an HLA gene, *Proc. Natl. Acad. Sci. U.S.A.* **79:**893–897.

Maxam, A., and Gilbert, W., 1977, Sequencing end-labeled DNA with base specific chemical cleavages, *Methods Enzymol.* **65:**499–599.

Miller, R. A., and Ruddle, F. H., 1977, Properties of teratocarcinoma–thymus somatic cell hybrids, *Somat. Cell Genet.* **3:**247–261.

Moore, K. W., Taylor, Sher, B., Sun, Y. H., Eakle, K. A., and Hood, L., 1982, DNA sequence of a gene encoding a Balb C mouse L^d transplantation antigen, *Science* **215:**679–682.

Nairn, R., Yamaga, K., and Nathenson, S. G., 1980, Biochemistry of the gene products from murine MHC mutants, *Annu. Rev. Genet.* **14:**241–277.

Orr, H. T., Lopez de Castro, J. A., Parham, P., Pleogh, H. L., and Strominger, J. L., 1979, Comparison of amino acid sequences of two human histocompatibility antigens HLA-A2 and HLA-B7: Location of putative alloantigenic sites, *Proc. Natl. Acad. Sci. U.S.A.* **76:**4395–4399.

Ploegh, H. L., Orr, H. T., and Strominger, J. L., 1981, Major histocompatibility antigens: The human (HLA-A, B, C) and murine (H2-K, D) class I molecules, *Cell* **24:**287–299.

Pober, J. S., and Strominger, J. L., 1981, Transglutaminase modifies the carboxy-terminal intracellular region of HLA-A and B-antigens, *Nature (London)* **289:**819–821.

Pober, J. S., Guild, B. C., Strominger, J. L., and Weatch, W. R., 1981, Purification of HLA-A$_2$ antigen, fluorescent labeling of its intracellular region, and demonstration of an interaction between fluorescently labeled HLA-A$_2$ antigen and lymphoblastoid cell cytoskeleton proteins *in vitro*, *Biochemistry* **20:**5625–5633.

Proudfoot, N. J., and Brownlee, G. G., 1976, 3'-Noncoding sequences in eucaryotic messenger RNA, *Nature (London)* **263:**211–214.

Shackelford, D. A., Mann, D. A., Van Rood, J. J., Ferrara, G. B., and Strominger, J. L., 1981, Human B cell alloantigens DC1, MT1 and LB12 are identical to each other but distinct from HLA-DR antigen, *Proc. Natl. Acad. Sci. U.S.A.* **7:**4566–4570.

Shaw, S., Kavathas, P., Pollack, M. S., Charmot, D., and Mawas, C., 1981, Family studies define a new histocompatibility locus, SB between HLA-DR and GLO, *Nature (London)* **299:**745–747.

Sood, A. K., Pereira, D. and Weissman, S. M., 1981, Isolation of a cDNA clone for human histocompatibility antigen HLA-B by use of an oligodeoxynucleotide primer, *Proc. Natl. Acad. Sci. U.S.A.* **78:**616–620.

Steinmetz, M., Moore, K. W., Frelinger, J. G., Shen, F., Boyse, E. A., and Hood, L., 1981, A pseudogene homologous to mouse transplantation antigens: Transplantation antigens are encoded by eight exons that correlate with protein domains, *Cell* **25:**683–692.

Tanigachi, N., Tosi, R., and Pressman, D., 1980, Molecular identification of human Ia antigens coded for by a gene locus closely linked to HLA-DR locus, *Immunogenetics* **10:**151–167.

Williams, R. A., Hart, D. N. J., Fabre, J. W., and Morris, P. J., 1980, Distribution and quantitation of HLA-A, B, C and DR(Ia) antigens on human kidney and other tissues, *Transplantation* **29:**274–279.

Zinkernagel, R. M., and Doherty, P. C., 1974, Restriction of *in vitro* T cell mediated cytotoxicity in lymphocyte choriomeningitis within a syngeneic or a semi-allogeneic system of mouse, *Science* **203:**516–521.

6

The Production of Transgenic Mice

JON W. GORDON and FRANK H. RUDDLE

Recombinant DNA technology allows the isolation of structural genes or various portions of their associated controlling elements. These cloned sequences can then be employed in gene-transfer experiments to study the regulation of gene expression. Such gene-transfer experiments using mammalian somatic cells in culture as recipients have produced a significant body of data relating to the control of gene function (Scangos and Ruddle, 1981). This experimental approach has the limitation, however, that cultured cells are unable to undergo many of the regulatory changes required for normal embryonic development. Thus, the genetic events responsible for processes such as organ determination and tissue differentiation are difficult if not impossible to examine. Our laboratory has sought to overcome this problem by developing a gene-transfer system for the intact organism. We have chosen the mouse as a model system.

Our initial concerns have been to demonstrate that: (1) cloned gene sequences can be introduced into the genomes of embryonic mice and (2) donor sequences can behave genetically like a resident mouse gene. Data to be reviewed below indicate that both these goals can be realized. In addition, preliminary findings will be described that indicate that donor sequences can

JON W. GORDON and FRANK H. RUDDLE • Department of Biology, Yale University, New Haven, Connecticut 06511.   Dr. Gordon's present address is Department of Obstetrics and Gynecology, Mount Sinai School of Medicine, New York, New York 10029.

act as mutagens, producing genetic lesions in the recipient that mimic naturally occurring developmental defects.

To study gene regulation from the onset of ontogeny, it is necessary to transfer genes at the earliest possible time in development. Accordingly, the one-celled mouse embryo was chosen as the recipient. DNA was transferred by microinjection into one of the pronuclei. Embryos were then reimplanted into the oviducts of pseudopregnant females and allowed to develop to term. DNA was extracted from the newborns, digested with restriction enzymes, electrophoresed on agarose, and tested by Southern blot hybridization (Blin and Stafford, 1976; Southern, 1975; Wahl *et al.*, 1979) for the presence of plasmid-derived sequences. The microinjection procedure is shown in Fig. 1. The recombinant plasmid chosen in these first experiments, pST6, contained the herpes virus thymidine kinase gene and segment of the simian virus 40 genome cloned into pBR322 (Gordon *et al.*, 1980). This plasmid has little sequence homology with the mouse genome, a fact that limited its cross-hybridization to the mouse DNA. The success of this gene-transfer strategy is shown by the Southern blot of DNA from the 48th mouse examined. Figure 2 shows DNA from this mouse and several of its littermates after digestion with *Bam*HI and hybridization against pST6. As this figure illustrates, positive animals can be readily distinguished from their negative counterparts. These data and those from the 73rd mouse examined (Gordon *et al.*, 1980) constitute a realization of the first goal—the successful transfer of exogenous genes into mouse embryos. This approach has subsequently been utilized by several other laboratories (Brinster *et al.*, 1981; Costantini and Lacy, 1981; Gordon and Ruddle, 1982; Wagner *et al.*, 1981a,b) to effect organismal gene transfer.

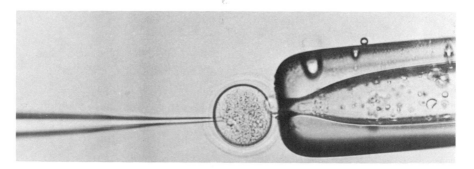

FIGURE 1.   Microinjection of the mouse embryo. The microneedle is inserted into the male pronucleus of this embryo.

For convenience, we have offered the term *transgenic* to describe mice into which foreign genetic material has been transferred (Gordon and Ruddle, 1981).

Two major criteria must be met to attain the second goal, that of demonstrating that transferred DNA can behave genetically like normal host genes. These criteria are: (1) integration into the host genome and (2) transmission of donor material through the germ line as a Mendelian trait. These two parameters are related insofar as Mendelian transmission indicates integration of donor material into a chromosome. However, because genetic information can be transmitted cytoplasmically, the inheritance of sequences from a parent does not by itself prove chromosomal integration.

To determine whether transgenic mice have integrated donor DNA into a chromosome, we have sought to demonstrate the acquisition of host-specific restriction-enzyme sites by the plasmid. This test has been used as a rigorous assessment of integration in studies involving the introduction of retroviruses into mouse embryos (Harbers *et al.*, 1981). In at least two transgenic mice thus far studied, integration of donor material could not be unequivocally

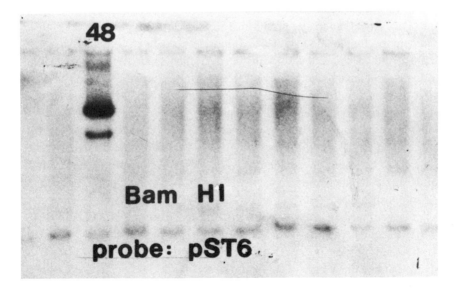

FIGURE 2.   Southern blot of DNA from several newborn mice injected at the pronuclear stage with the recombinant plasmid pST6. The DNA was digested with *Bam*HI and probed with radiolabeled pST6. The DNA from mouse 48 manifests marked hybridization to the probe; the littermates are negative.

documented (Gordon *et al.*, 1980; Gordon and Ruddle, 1982). However, in three other cases, restriction fragments of the injected plasmid did require sites for enzymes specific for host DNA. Two animals were derived from injections of pST6, which has no recognition sites for the restriction enzyme *Xba*I. Double digests of DNA from these mice (first with *Bam*HI and then with *Xba*I) show that several kilobases of DNA were removed from one *Bam*HI fragment in each mouse after digestion with *Xba*I (Gordon and Ruddle, 1981, 1982). The third example involved injections of plasmid pIf, which comprises the coding sequence of a human leukocyte interferon cloned in pBR322. Like pST6, pIf contains no restriction sites for *Xba*I. Partial digests with *Xba*I of spleen DNA from transgenic mouse No. If-4 produced multiple bands, all of higher molecular weight than a single 13.5-kilobase band produced by complete digestion with this enzyme (Gordon and Ruddle, 1981). This result shows an association of pIf DNA in If-4 with long host sequences containing numerous sites for *Xba*I. Thus, this transgenic mouse fulfills the criteria for integration of donor DNA. In sum, we have produced several transgenic mice in which restriction-enzyme analyses strongly indicate integration of the microinjected plasmid.

If-4 has been additionally tested for transmission of pIf through the germ line. This animal, a female, produced 15 offspring, 9 of which were positive (Gordon and Ruddle, 1981). It was important to the germ-line transmission test that the specific alterations in the restriction map of pIf that occurred during integration into the genome of If-4 were also demonstrable in offspring. Inheritance of the altered plasmid would indicate that the pIf DNA in offspring could have come only from If-4. We therefore digested the DNA of 6 progeny with four restriction enzymes and compared the patterns of fragments obtained with that of If-4—they were indistinguishable (Gordon and Ruddle, 1981). Particularly compelling were the results of partial *Xba*I digestion, after which the multiple fragments generated in If-4 and 1 of its progeny were all of indistinguishable size (Gordon and Ruddle, 1981, 1982).

Further breeding of the If-4 line gave a pattern of transmission indicative of Mendelian inheritance. Thus far, 121 animals have been analyzed; 62 were positive and 59 were negative. These numbers conform best to a pattern of Mendelian inheritance after integration of the plasmid into one chromosomal homologue. However, the pIf-derived material has been transmitted only through females, because males carrying this DNA are uniformly sterile. This observation will be expanded on below, where the use of transgenosis to produce developmental lesions is discussed. Suffice it to say at this point that the pIf DNA in the If-4 line cannot be said to behave as a typical genetic element of the mouse insofar as it cannot be transmitted through males.

Transmission through both sexes has been observed in an additional line of transgenic mice. The original transformant, If(gen.) 20.05, was produced by pronuclear microinjection of a genomic human interferon gene cloned in the *Bam*HI site of pBR322 (Mantei and Weissman, 1982). Restriction analysis indicates that this animal has incorporated a large concatameric derivative of the recombinant plasmid; such concatamers have been observed in similar experiments from other laboratories (Costantini and Lacy, 1981; Wagner *et al.*, 1981b). Of 38 progeny of If(gen.) 20.05 thus far tested, 20 are positive and 18 are negative. Several of these positive progeny, both male and female, have been subsequently crossed and have transmitted the pCHR-35 sequence. Figure 3 depicts germ-line transmission through If(gen.) 20.05 and one of its male progeny. Two positive females have also been crossed with a positive male to yield 13 offspring, 76.9% of which carry the pCHR-35 concatamer. This latter number corresponds closely to the 75% expected from Mendelian transmission of a trait through two heterozygous carriers. These breeding data indicate that the transferred sequence is transmitted in a manner entirely analogous to a Mendelian trait.

These data thus satisfy important prerequisites for the development of an organismal gene-transfer system capable of providing data representative of the activity of normal genetic elements within the mouse. The If-4 line of mice, however, is not normal, in that males are unable to transmit the pIf sequence. Our preliminary analysis of this anomalous transmission indicates another potential use of the transgenic mouse system—the ability to interrupt sequences with noncoding donor DNA and thereby induce specific developmental lesions.

The female transgenic mouse If-4 has been extensively bred to produce a colony of animals carrying the modified recombinant plasmid pIf. Of the 44 males analyzed thus far, 26 are positive and 18 are negative. In all 26 positive cases, the males are sterile. These animals have normal secondary sex characteristics and mate normally, producing vaginal plugs in estrous females. However, no spermatozoa can be recovered from the reproductive tracts of females thus mated, and all oocytes are unfertilized. The sex ducts of these males are normal in appearance, but spermatozoa in the caudae epidydimis are few or absent. Moreover, the testes are reduced in size significantly when compared with age-matched controls. Preliminary histological analysis indicates that spermatogenesis is interrupted in the transgenic males.

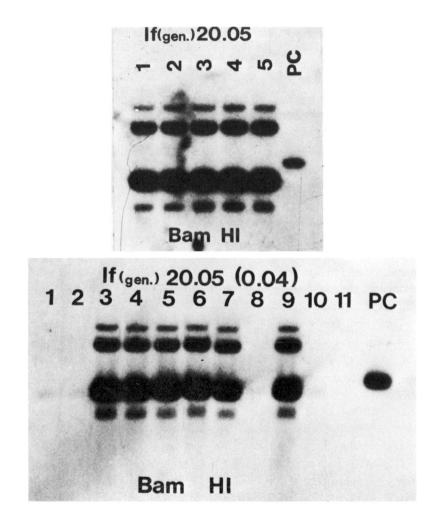

FIGURE 3. (Top) Five progeny of the transgenic female mouse If(gen.) 20.05. The two intense bands of intermediate size represent the interferon fragment and the cloning vector, pBR322. (PC) Positive control, which in this case was composed of *Bam*HI-digested plf DNA (see the text). (Bottom) Eleven progeny of the 4th offspring of If(gen.) 20.05. This animal, a male, transmitted the pCHR-35 sequence to 6 of the 11 animals shown. This result demonstrated Mendelian transmission of donor DNA through both sexes.

Experiments are currently under way to determine the precise cause of the association between the presence of pIf DNA and sterility. Production of human interferon is most likely not the cause, since the coding sequence cloned into this plasmid lacks promoters, and restriction-enzyme analysis indicates a disruption of the interferon structural gene in the transgenic mice. It is more likely, therefore, that the donor DNA has altered the host genome so as to impair spermatogenesis.

The advantage of the transgenic mouse system is that important host sequences such as those disrupted by pIf DNA in this line can be recovered from cloned genomic libraries of transgenic mouse DNA using the plasmid as a probe. The sequences can then be subjected to molecular analysis with the reasonable expectation that such analysis will provide insight into the molecular mechanisms underlying such developmental changes. We have undertaken to clone pIf sequences from genomic DNA libraries of positive mice from the If-4 line. One of three *Eco*RI fragments, the largest, has been cloned and shown to contain a junction between pIf and host DNA (McClelland *et al.*, 1984). This finding formally demonstrates covalent linkage of donor and recipient DNA and will shortly allow a direct examination of the integrated plasmid and its surrounding material. It is anticipated that this analysis will explain the phenomenon of male sterility in males of this line.

In conclusion, our laboratory has developed a gene-transfer system for the intact mammalian organism. Initial evaluation indicates that DNA transferred into mice can become integrated into the host genome and transmitted to progeny through both sexes. At this point, therefore, transgenic mice appear to have significant potential for molecular analysis of mammalian development. In addition, the transgenic mouse may serve as a system for creation of genetic lesions affecting development and the subsequent analysis of the molecular basis of these lesions.

REFERENCES

Blin, N., and Stafford, D. W., 1976, A general method for isolation of high molecular weight DNA from eukaryotes, *Nucleic Acids Res.* 3:2303–2308.

Brinster, R. L., Chen, H. Y., Trumbauer, M., Senear, A. W., Warren, R., and Palmiter, R. D., 1981, Somatic expression of herpes thymidine kinase in mice following injection of a fusion gene into eggs, *Cell* 27:223–231.

Costantini, F., and Lacy, E., 1981, Introduction of a rabbit beta-globin gene into the mouse germ line, *Nature (London)* 294:92–94.

Gordon, J. W., and Ruddle, F. H., 1981, Integration and stable germ line transmission of genes injected into mouse pronuclei, *Science* 214:1244–1246.

Gordon, J. W., and Ruddle, F. H., 1982, Germ line transmission in transgenic mice, in: *Embryonic Development*, Part B, *Cellular Aspects* (M. M. Burger and R. Weber, eds.), Alan R. Liss, New York, pp. 111–124.

Gordon, J. W., Scangos, G. A., Plotkin, D. J., Barbosa, J. A., and Ruddle, F. H., 1980, Genetic transformation of mouse embryos by microinjection of purified DNA, *Proc. Natl. Acad. Sci. U.S.A.* **77:**7380–7384.

Harbers, K., Jahner, D., and Jaenisch, R., 1981, Microinjection of cloned retroviral genomes into mouse zygotes: Integration and expression in the animal, *Nature (London)* **293:**540–542.

Mantei, N., and Weissman, C., 1982, Controlled transcription of a human α-interferon gene introduced into mouse L cells, *Nature (London)* **297:**128–132.

McClelland, A., Gordon, J. W., and Ruddle, F. H., 1984, Molecular cloning of integrated DNA from a transgenic mouse (in preparation).

Scangos, G. A., and Ruddle, F. H., 1981, Mechanisms and applications of DNA-mediated gene transfer in mammalian cells—a review, *Gene* **14:**1–10.

Southern, E. M., 1975, Detection of specific sequences among DNA fragments separated by gel electrophoresis, *J. Mol. Biol.* **98:**503–517.

Wagner, T. E., Hoppe, P. C., Jollick, J. D., Scholl, D. R., Hodinka, R. L., and Gault, J. B., 1981a, Microinjection of a rabbit beta-globin gene into zygotes and its subsequent expression in adult mice and their offspring, *Proc. Natl. Acad. Sci. U.S.A.* **78:**6376–6380.

Wagner, E. F., Stewart, T. A., and Mintz, B., 1981b, The human beta-globin gene and a functional viral thymidine kinase gene in developing mice, *Proc. Natl. Acad. Sci. U.S.A.* **78:**5016–5020.

Wahl, G. M., Stern, M., and Stark, G. R., 1979, Efficient transfer of large DNA fragments from agarose gels to dioxybenzoyloxymethyl-paper and rapid hybridization using dextran sulfate, *Proc. Natl. Acad. Sci. U.S.A.* **76:**3683–3687.

7

Human and Mouse Globin-Gene Sequences Introduced into Mice by Microinjection of Fertilized Mouse Eggs

R. K. HUMPHRIES, P. BERG, J. DiPIETRO,
S. BERNSTEIN, A. BAUR, A. NIENHUIS, and
W. F. ANDERSON

1. INTRODUCTION

The globin gene families provide an interesting model of regulated gene expression during both ontogenic development and erythroid differentiation (reviewed in Stamatoyannopoulos and Nienhuis, 1981). From sequence data and functional analysis of cloned globin genes, some regions (such as the consensus CCAAT and TATA blocks 5' to coding sequences) have been identified as important for influencing the rate and accuracy of transcript initiation (Efstratiadis *et al.*, 1980; Lauer *et al.*, 1980; Dierks *et al.*, 1981; Mellon *et al.*, 1981; Grosveld *et al.*, 1982).

R. K. HUMPHRIES, A. BAUR, and A. NIENHUIS ● Clinical Hematology Branch, National Heart, Lung, and Blood Institute, National Institutes of Health, Bethesda, Maryland 20205. P. BERG, J. DiPIETRO, S. BERNSTEIN, and W. F. ANDERSON ● Laboratory of Molecular Hematology, National Heart, Lung, and Blood Institute, National Institutes of Health, Bethesda, Maryland 20205.

Much less is known about sequences responsible for tissue- and stage-specific expression of genes during ontogeny. One potentially powerful model for identifying such regulatory sequences may be provided through analysis of defined DNA segments transferred to mice by microinjection at the zygote stage of development. Integration into the mouse genome, and subsequent germ-line transmission, has now been demonstrated by several groups for a variety of cloned fragments including the rabbit β-globin gene (T. E. Wagner *et al.*, 1981; Costantini and Lacy, 1981), herpes simplex thymidine kinase gene (Gordon *et al.*, 1980; E. F. Wagner *et al.*, 1981), fusion genes of the metallothionein promoter and the thymidine kinase gene (Brinster *et al.*, 1981), and a complementary DNA (cDNA) human leukocyte interferon (Gordon and Ruddle, 1981). Low but detectable levels of expression in adult mice have been reported for genes coupled to the metallothionein promoter (Brinster *et al.*, 1981; Palmiter *et al.*, 1982). T. E. Wagner *et al.* (1981) have reported expression of the rabbit β-globin gene in mouse erythrocytes, whereas Costantini and Lacy (1981) detected no expression of this gene in their studies.

We have begun to use the technology of gene transfer to mouse eggs by microinjection to try to identify regulatory sequences required for human and mouse globin-gene expression. In this chapter, we describe results of our initial studies in which, in one instance, we have obtained germ-line transmission and integration of the human δ- and β-globin genes and, in another instance, of plasmid sequences containing the mouse β-major globin promoter coupled to a prokaryotic enzyme-coding sequence.

2. MATERIALS AND METHODS

2.1. Mice

All mice, with the exception of NIH Swiss, were obtained from Jackson Laboratories. NIH Swiss were obtained from the stock of NIH. Fertilized eggs at the pronuclear stage were harvested on the morning of vaginal plugging (day 1) from C57B1/6 females mated to LT/SV males. Eggs were stored and cultured in modified Whittens medium (Hoppe and Pitts, 1973), kept under paraffin oil equilibrated with the same medium, and gassed with 5% CO_2, 5% O_2, and 90% N_2 (T. E. Wagner *et al.*, 1981). Pseudopregnant recipient mice were obtained by mating B6D2F1 females with vasectomized NIH Swiss males.

2.2. Microinjection

Eggs were injected with approximately 10 pl DNA solution essentially as described by T. E. Wagner *et al.* (1981). Eggs were cultured for 4 days, and those reaching the morula stage were transferred to the uteri of B6D2F1 day 3 pseudopregnant females. Pregnancies were allowed to go to term, and mice were born naturally or by caesarean section.

2.3. DNA for Injection

The recombinant phage λHβG1, containing both the human δ- and β-globin genes, was kindly provided by Dr. Tom Maniatis and co-workers (see Fig. 1A) (Lawn *et al.*, 1978). The recombinant plasmid pPB22 contains the

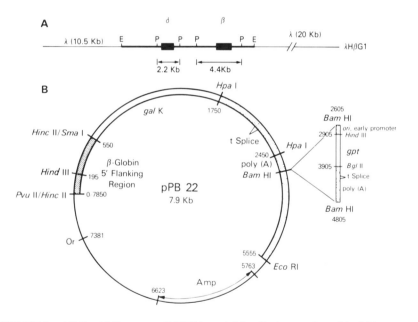

FIGURE 1.   (A) Restriction map of λHβG1 containing the human δ- and β-globin genes. The dark boxes indicate the positions of the human δ- and β-globin structural genes including introns. Restriction enzymes: (E) *Eco*RI; (P) *Pst*I. Indicated below the map are the fragments released by *Pst*I and detected with a ^{32}P-labeled globin cDNA probe prepared from reticulocyte messenger RNA. (B) Restriction map of pPB22. Each region is labeled. The t splice, polyadenylic acid [poly(A)] region, and *"ori,* early promoter" regions are all from SV40. The details of this construction have been published elsewhere (Berg *et al.,* 1983).

mouse β-major globin gene promoter and the simian virus 40 (SV40) promoter fused to coding regions of the *Escherichia coli* galactokinase gene and the *E. coli gpt* gene, respectively (see Fig. 1B). Details of the construction of this plasmid will be described elsewhere (Berg *et al.*, 1983). DNA from λHβG1 or pPB22 was injected in closed circular form.

2.4. DNA Analysis

Animals were screened for the presence of injected sequences using DNA isolated from spleens of adult animals or portions of tails removed at 4 weeks of age. DNA from spleens was obtained by lysis of cells in guanidine hydrochloride and cesium chloride centrifugation (Glisin *et al.*, 1974; Kantor *et al.*, 1980). Tail DNA was isolated and spot blots performed essentially as described (Palmiter *et al.*, 1982). Southern (1975) blot analysis following restriction-enzyme digestion was performed in standard fashion. ^{32}P-labeled probes were prepared by synthesis of cDNA from human reticulocyte RNA (Benz *et al.*, 1977) or by nick translation of DNA fragments (as described by the Bethesda Research Laboratory in its analysis kit).

3. RESULTS

3.1. Transfer of Human Globin-Gene Sequences

In the first study, we injected DNA isolated from the bacteriophage λHβG1, which contains both the human δ- and β-globin genes (Lawn *et al.*, 1978) (see Fig. 1A). Eggs were injected into the male pronucleus, cultured to the morula stage, and transferred to pseudopregnant mothers. Six live offspring were obtained. At approximately 8 weeks of age, these mice were treated with phenylhydrazine. DNA and RNA were isolated from the erythroid spleen. The result of Southern blot analysis of *Pst*I-restricted DNA is shown in Fig. 2. Human globin-gene sequences were detected by a ^{32}P-labeled cDNA probe prepared from human reticulocyte RNA (Benz *et al.*, 1977). Some cross-hybridization to mouse globin sequences is evident in all mouse DNA samples. In one animal (lane 5), intense hybridization signals were observed at positions of 4.4 and 2.3 kb corresponding to the fragment sizes encompassing the human β- and δ-globin genes, respectively. From comparison of the intensity of this signal to lanes containing known amounts of λHβG1, we

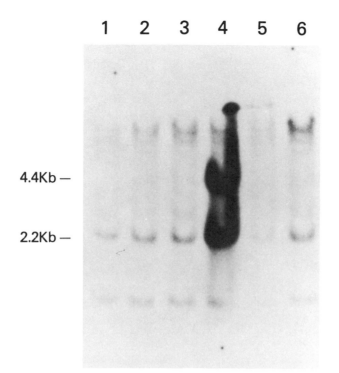

FIGURE 2. Demonstration of human globin-gene sequences in mouse DNA by Southern blot analysis. DNA was purified from spleen tissue of a control mouse (lane 6) and mice born following injection of zygotes with λHβG1 DNA (lanes 1–5). Following restriction of 30 μg DNA with $PstI$, DNA was electrophoresed on 1% agarose gel, transferred to nitrocellulose paper, and hybridized to an $\alpha^{32}P$-labeled globin cDNA probe prepared from human reticulocyte RNA. The intense 4.4-kilobase (kb) and 2.3-kb bands in lane 5 represent the human β- and δ-globin genes, respectively. Faint bands seen in all lanes represent cross-hybridization of the cDNA probe to endogenous mouse globin-gene sequences.

estimate that the copy number in this animal is greater than 20. Repeated attempts to mate this male resulted in pregnancies that arrested approximately in midterm with resorption of the fetuses. At approximately 1 year of age, however, this mouse sired two offspring, one of which was positive for the human δ- and β-globin-gene sequences. In RNA analysis of the primary animal, no β- or δ-globin messenger RNA was detectable either by liquid hybridization using ^{32}P-labeled globin cDNA or by sensitive S_1 analysis per-

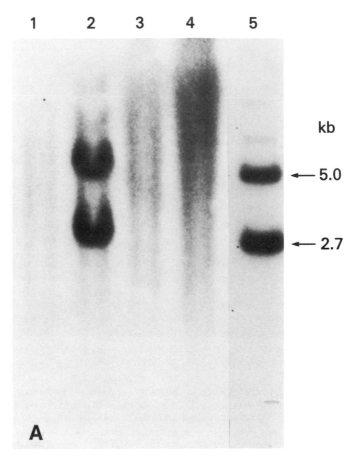

FIGURE 3. Southern blot analysis of mice born following injection with plasmid pPB22. (A) DNA (30 μg) from spleens of four mice born following microinjection of pPB22 DNA (Lanes 1–4) and 10 ng purified pPB22 DNA (lane 5) was digested with *Hind*III, electrophoresed on 1% agarose gel, and transferred to nitrocellulose paper. Plasmid sequences were visualized following hybridization to a nick-translated $\alpha^{32}P$-labeled probe

formed using uniformly labeled single-stranded globin DNA probes (Ley *et al.*, 1982).

3.2. Transfer of Mouse Globin-Gene Sequences

In a second series of injections, we used a recombinant plasmid, pPB22, that contains both the mouse β-major globin gene promoter and SV40 pro-

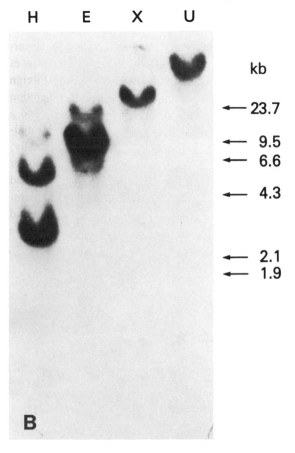

H   E   X   U

kb

◄— 23.7

◄— 9.5
◄— 6.6

◄— 4.3

◄— 2.1
◄— 1.9

B

made from a DNA fragment containing galactokinase and SV40 sequences. (B) Further analyses of DNA from the positive mouse [lane 2 in (A)]. Spleen DNA (30 μg) from the same mouse as in lane 2 above was digested with *Hind*III (H), *Eco*RI (E), *Xba*I (X), or no enzyme (U). DNA was blotted and hybridized as in (A). The positions of λ *Hind*III cut marker fragments are indicated.

moters; these promoters are linked to the coding region of the prokaryotic *gal K* gene or *gpt* gene, respectively (Fig. 1B). The rationale for use of this plasmid was twofold: first, the effect of the species of origin of the promoter on expression could be tested; second, gene expression could be sensitively assayed by a specific enzyme assay. Of 12 mice born, 2 were positive for injected DNA sequences as assessed by Southern blot analysis such as shown in Fig. 3A. One positive animal died shortly after birth and was not further analyzed. Further restriction-enzyme analysis of DNA obtained from the surviving animal is shown in Fig. 3B. In uncut DNA, hybridization sequences were present in the high-molecular-weight fraction, suggesting that the foreign sequences were integrated into host genomic DNA. Analysis with restriction enzyme *Xba*I, which has no sites in the plasmid DNA injected, released a single fragment of approximately 30 kb. This observation suggests that multiple copies of the injected DNA were integrated as a concatamer. Finally, restriction analysis with *Eco*RI yielded a pattern identical to that obtained for the plasmid DNA, indicating that no major deletions or rearrangements had occurred. Spot blot analysis of offspring of this mouse (Fig. 4) revealed germline transmission of the injected DNA sequences through two generations and to approximately 50% of offspring (Fig. 5). Neither prokaryotic galactokinase

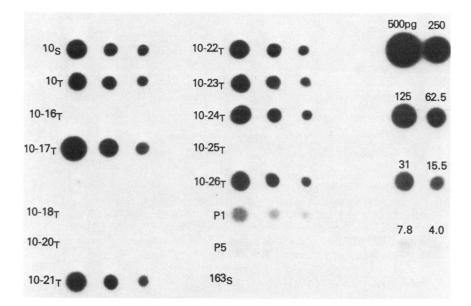

FIGURE 4.   Demonstration of germ-line transmission of microinjected DNA containing the cloned β-globin-promoted *gal* K gene and SV40-promoted *gpt* gene. The positive primary animal (No. 10), ascertained from the data in Fig. 3, was mated to a negative male. DNA, 5, 2.5, or 1.0 μg, isolated from the primary animal's spleen (10_S) or tail (10_T) and DNA from tails of offspring were spotted on nitrocellulose paper. Hybridization was to a probe as described in Fig. 3. Animals P1 and P5 are from an unrelated series of injections. (163_S) DNA from a noninjected mouse. Known quantities of the original plasmid DNA (right side of figure) were spotted for estimation of copy number.

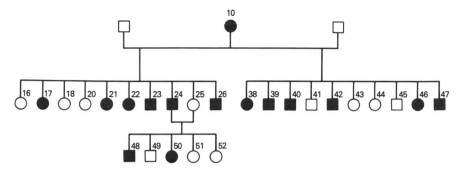

FIGURE 5. Pedigree of transgenic mouse. The positive mouse (No. 10) is described in the captions of Figs. 3 and 4. Numbers refer to mouse ear tags of offspring.

nor *gpt* enzyme activity was detectable in red blood cells or spleen cells taken from the primary animal after phenylhydrazine treatment. Gene expression in offspring is currently under analysis.

4. DISCUSSION

These studies have demonstrated germ-line integration and transmission in mice of cloned DNA sequences containing the human δ- and β-globin genes and the mouse β-major globin promoter linked to the *E. coli gal* K gene. These results corroborate those of other groups that gene transfer via microinjection of fertilized mouse eggs can occur with relatively high frequency (15% in the studies presented herein). No evidence of expression of the transferred DNA was demonstrated in erythroid tissue of the primary animals. Examination of other tissues and in secondary animals is obviously important in view of the reported variation in expression seen by others (Palmiter *et al.*, 1982). It should also be of interest to determine whether the lack of expression is due to *de novo* methylation of injected DNA (Palmiter *et al.*, 1982; Jahner *et al.*, 1982). Further experiments are necessary to determine whether gene expression can be obtained using larger cloned genomic fragments or other gene promoters or both. Such studies may provide the basis for obtaining regulated gene expression of transferred DNA.

REFERENCES

Benz, E. J., Jr., Geist, C. E., Steggles, A. W., Barker, J. E., and Nienhuis, A. W., 1977, Hemoglobin switching in sheep and goats: Preparation and characterization of complementary DNAs specific for the alpha, beta, and gamma globin messenger RNAs of sheep, *J. Biol. Chem.* **252:**1908–1916.

Berg, P. E., Yu, J.-K., Popovic, Z., Schumperli, D., Johansen, H., Rosenberg, M., and Anderson, W. F., 1983, Differential activation of the mouse β-globin promoter by enhancers, *Mol. Cell. Biol.* **3:**1246–1254.

Brinster, R. L., Chen, H. Y., Trumbauer, M., Senear, A. W., Warren, R., and Palmiter, R. D., 1981, Somatic expression of herpes thymidine kinase in mice following injection of fusion gene into mice, *Cell* **27:**223–231.

Costantini, F., and Lacy, E., 1981, Introduction of a rabbit β-globin gene into the mouse germ line, *Nature (London)* **294:**92–94.

Dierks, P., van Ooyen, A., Mantei, N., and Weissman, C., 1981, DNA sequences preceding the rabbit β-globin gene are required for formation in mouse L cells of β-globin RNA with the correct 5′ terminus, *Proc. Natl. Acad. Sci. U.S.A.* **78:**1411–1415.

Efstratiadis, A., Posakony, J. W., Maniatis, T., Lawn, R. M., O'Connell, C., Spritz, R. A., DeRiel, J. K., Forget, B. G., Weissman, S. M., Slightor, J. L., Blechl, A., Smithies, O., Baralle, F. E., Shoulders, C. C., and Proudfoot, N. J., 1980, The structure and evolution of the human β-globin gene family, *Cell* **21:**653–668.

Glisin, V., Crkvenjakov, R., and Byus, C., 1974, Ribonucleic acid isolated by cesium chloride centrifugation, *Biochemistry* **13:**2633–2637.

Gordon, J. W., and Ruddle, F. H., 1981, Integration and stable germ line transmission of genes injected into mouse pronuclei, *Science* **214:**1244–1246.

Gordon, J. W., Scangos, G. A., Plotkin, O. J., Barbosa, J. A., and Ruddle, F. H., 1980, Genetic transformation of mouse embryos by microinjection of purified DNA, *Proc. Natl. Acad. Sci. U.S.A.* **77:**7380–7384.

Grosveld, G. C., deBoer, E., Shewmaker, C. K., and Flavell, R. A., 1982, DNA sequences necessary for transcription of the rabbit β-globin gene *in vivo*, *Nature (London)* **295:**120–126.

Hoppe, P. C., and Pitts, S., 1973, Fertilization *in vitro* and development of mouse ova, *Biol. Reprod.* **8:**420–426.

Jahner, D., Stuhlmann, H., Stewart, C. L., Harbers, K., Lohler, J., Simon, I., and Jaenisch, R., 1982, *De novo* methylation and expression of retroviral genomes during mouse embryogenesis, *Nature (London)* **298:**623–628.

Kantor, J. A., Turner, P. H., and Nienhuis, A. W., 1980, Beta thalassemia: Mutations which affect processing of the β-globin mRNA precursor, *Cell* **21:**149–157.

Lauer, J., Shen, C. K. J., and Maniatis, T., 1980, The chromosomal arrangements of the human α-like globin genes: Sequence homology and α-globin gene deletions, *Cell* **20:**119–130.

Lawn, R. M., Fritsch, E. F., Parker, R. C., Blake, G., and Maniatis, T., 1978, The isolation and characterization of linked δ and β globin genes from a cloned library of human DNA, *Cell* **21:**1157–1174.

Ley, T. J., Anagnou, N. P., Guglielmina, P., and Nienhuis, A. W., 1982, RNA processing errors in patients with β-thalassemia, *Proc. Natl. Acad. Sci. U.S.A.* **79:**4775–4779.

Mellon, P., Parker, V., Gluzman, Y., and Maniatis, T., 1981, Identification of DNA sequences required for transcription of the human α1-globin gene in a new SV40 host–vector system, *Cell* **27:**279–288.

Palmiter, R. D., Chen, H. Y., and Brinster, R. L., 1982, Differential expression of metallothionein–thymidine kinase fusion genes in transgenic mice and their offspring, *Cell* **29:**701–710.

Southern, E. M., 1975, Detection of specific sequences among DNA fragments separated by agarose gel electrophoresis, *J. Mol. Biol.* **98**:503–515.

Stamatoyannopoulos, G., and Nienhuis, A. W. (eds.), 1981, *Organization and Expression of Globin Genes,* Alan R. Liss, New York.

Wagner, E. F., Stewart, T. A., and Mintz, B., 1981, The human β-globin gene and a functional viral thymidine kinase gene in developing mice, *Proc. Natl. Acad. Sci. U.S.A.* **78**:5016–5020.

Wagner, T. E., Hoppe, P. C., Jollick, J. D., Scholl, D. R., Holinka, R. L., and Gault, J. B., 1981, A preliminary report of the microinjection of a rabbit β globin gene into zygotes and its subsequent expression in the adult mice, *Proc. Natl. Acad. Sci. U.S.A.* **78**:6376–6380.

8

A Novel System Using the Expression of Chloramphenicol Acetyltransferase in Eukaryotic Cells Allows the Quantitative Study of Promoter Elements

CORNELIA GORMAN, LAIMOMIS LAIMONS, GLENN T. MERLINO, PETER GRUSS, GEORGE KHOURY, and BRUCE HOWARD

1. INTRODUCTION

As the number of isolated putative eukaryotic promoter sequences has increased, so has the need for an accurate means of measuring the function of these sequences. The *in vitro* transcription systems developed by Manley *et al.* (1980) and Weil *et al.* (1979) offer one approach. However, it is becoming clear that the *in vitro* transcription systems may respond to different regulatory

CORNELIA GORMAN, GLENN T. MERLINO, and BRUCE HOWARD ● Laboratory of Molecular Biology, National Cancer Institute, National Institutes of Health, Bethesda, Maryland 20205.   LAIMOMIS LAIMONS, PETER GRUSS, and GEORGE KHOURY ● Laboratory of Molecular Virology, National Cancer Institute, National Institutes of Health, Bethesda, Maryland 20205.

signals and thus do not afford the ideal system for the study of *in vivo* transcriptional control (Benoist and Chambon, 1980). The study of promoters after introduction into tissue-culture cells is crucial.

To study regulation of a cloned gene, it is often desirable to reintroduce that cloned gene into the homologous cell type. However, in the homologous cell type, the problem of distinguishing expression of the cloned and endogenous genes is most pronounced. Although RNA levels represent the most definitive gauge of promoter activity, it can be difficult to quantitate these levels accurately unless a given promoter is particularly strong, or it can be difficult to distinguish the RNA of interest from the basal levels of RNA within the cell. Even in systems in which the plasmid RNA can be distinguished from the endogenous RNA (Mellon *et al.,* 1981; de Villers and Schaffner, 1981), it has been found that the levels of RNA are too low to detect unless enhancer sequences are included in the plasmid.

One solution to this problem is to combine the transcription start site or putative regulatory region(s), or both, of a cloned gene of interest with a second gene segment that provides an easily assayable and readily distinguished function. In selecting an enzymatic function to monitor promoter function, there are several relevant considerations. Most important is the absence of any corresponding endogenous enzymatic activity from any host cell. This greatly increases the sensitivity of detection. Second, there should be no interference from other enzymatic activities that could compete for utilization of substrate or cofactors. Third, the assay should be rapid and reproducible. With these considerations in mind, we have developed a series of recombinants in which the enzyme chloramphenicol acetyltransferase (CAT) is used to study promoter elements in eukaryotic cells (Gorman *et al.,* 1982a).

The CAT system has been used to study a variety of promoters including the entire simian virus 40 (SV40) early region promoter (Gorman *et al.,* 1982a), the chick αI (II) collagen promoter, the herpes simplex virus (HSV) thymidine kinase (TK) promoter, and the Rous sarcoma virus (RSV) long terminal repeat (LTR) as a promoter (Gorman *et al.,* 1982b). Additionally, the enhancer function of the 72-base-pair (bp) repeats of SV40 (Gruss *et al.,* 1981; Benoist and Chambon, 1981; Moreau *et al.,* 1981) has been quantitated (Gorman *et al.,* 1982a) and contrasted with the enhancer function of the murine sarcoma virus (MSV) repeats (Levinson *et al.,* 1982; Laimons *et al.,* 1982) using the CAT system. Here, we will summarize these data as well as discuss the practical aspects of the use of the CAT system.

2. CONSTRUCTION OF PLASMID pSV2cat

The *Escherichia coli* transposable element Tn9 (Scott, 1973), which confers resistance to the antibiotic chloramphenicol, consists of a 1102-bp *CAT* gene flanked by two 768-bp IS1 elements. A 773-bp fragment lacking the bacterial CAT promoter sequences was generated by a complete digestion of pBR322-Tn9 (J. L. Rosner, NIH) with *Taq*I. This fragment was modified by incubation with DNA polymerase I to create blunt ends and addition of mixed *Hind*III and *Bam*HI synthetic oligonucleotide linkers. This 773-bp fragment was joined with the prokaryotic/eukaryotic vector pSV2 (B. Howard

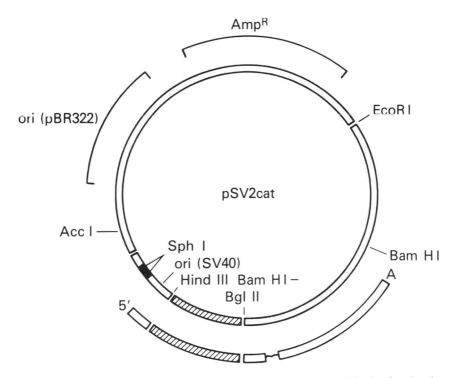

FIGURE 1.   Illustration of pSV2cat. Plasmid pSV2cat contains the following functional elements (counterclockwise from 12 o'clock on the circular map): The Amp^R cistron and the origin of replication from pBR322 (2296 bp); segments containing the SV40 early promoter region (400 bp), the CAT coding region from Tn9 (773 bp), the SV40 small t intron (610 bp), and the SV40 early region poly(A) addition site (988 bp).

and P. Berg, unpublished results), which consists of the origin of replication and ampillicin resistance genes of pBR322 (*Pvu*II–*Eco*RI, 2295 bp), an SV40 early region transcription unit (400 bp) modified to accept foreign polypeptide coding sequences between *Hind*III and *Bgl*II sites, the small t intron (610 bp), and the SV40 early region polyadenylic acid [poly(A)] addition site (988 bp). In the recombinant, pSV2cat (Fig. 1) (Gorman *et al.*, 1982a), the 773-bp bacterial insert consists of a 29-bp 5′ untranslated segment, the CAT coding sequence, and 86 bp 3′ to the translation stop codon. Since the promoter and 32 bp corresponding to the 5′ untranslated region of the CAT messenger RNA (mRNA) are deleted from the bacterial insert in pSV2cat, *E. coli* that carry this recombinant are sensitive to chloramphenicol.

3. EUKARYOTIC CELL TRANSFECTIONS

One important use of the CAT system has been the optimization of transfection procedures. The conditions given here yield maximum levels of CAT activity in monkey kidney CV-1 cells. It is important to use plasmid DNA free of chromosomal DNA. This is accomplished by cesium chloride–ethidium bromide equilibrium gradient centrifugation (Radloff *et al.*, 1967). DNAs that had been nicked randomly give unreproducible results and therefore are not used. The density of the cells at the time of transfection has been found to greatly affect the levels of transient expression. It is important that cells be in log phase of growth when the DNA is added to the cells. This is done by splitting the cells the day prior to transfection. At 3 hr before the addition of DNA, cells are refed with fresh medium containing 10% fetal calf serum. Calcium phosphate–DNA precipitates are prepared according to Graham and van der Eb (1978). Care is taken to prepare very fine precipitates; this is accomplished by mixing the DNA–CaCl$_2$ and HEPES-buffered sodium phosphate solutions under a gentle stream of nitrogen. Precipitates are allowed to stand 20 min without agitation, then added to tissue-culture cells. Initially, the amount of DNA added was varied between 1 and 25 μg/plate to determine the linear range of uptake and expression of the recombinants. Maximum CAT expression is obtained with 10 μg DNA/100-mm plate. At 4 hr after the addition of DNA, cells are treated with a glycerol shock (Parker and Stark, 1979). The length of this treatment may vary with cell type; for instance, CV-1 cells are treated for 3 min, while primary chick embryo fibroblasts (CEF) are treated for 30 sec. Cells are harvested between 36 and 48 hr; this has been determined to be optimum for transient CAT expression. Transient CAT expression has been monitored in monkey CV-1, mouse NIH/3T3, Quail 35, mouse L, HeLa, and Chinese hamster ovary cells, and in CEF with no background.

4. ASSAY OF CHLORAMPHENICOL ACETYLTRANSFERASE ACTIVITY IN EUKARYOTIC CELLS

CAT inactivates chloramphenicol by formation of mono- and diacetylated derivatives (Shaw, 1967). A number of assays have been developed to measure this activity (Shaw, 1967, 1975; Robinson *et al.*, 1978; Smith and Smith, 1978); we have adopted the assay described by Cohen *et al.* (1980) and Shaw and Brodsky (1978), in which acetylation of chloramphenicol is measured by silica gel thin-layer chromatography (TLC). This assay for CAT is very highly specific, easily separating the parent and mono- and diacetylated forms of chloramphenicol. Perhaps more important is the fact that this assay is very sensitive. Using purified CAT protein (gift of W. Shaw, Leicester, England), we have determined that as little as 0.4 pg enzyme is readily detectable. Cell extracts are made by sonicating washed, pelleted cells in 100 μl 0.25 M Tris (hydroxymethyl)aminomethane (Tris HCl), pH 7.8. Following a 15-min spin in an Eppendorf microfuge at 4°C, the supernatants are assayed for enzyme activity. Assays contain in a final volume of 180 μl: 100 μl 0.25 M Tris HCl, pH 7.5; 20 μl cell extract; 1 μCi [^{14}C]chloramphenicol (50 mCi/mmole, New England Nuclear); and 20 μl 4 mM acetyl-CoA. Controls contain CAT (0.01 U, P. L. Biochemicals) instead of cell extract. Assays are incubated for 30 min at 37°C. The actual amount of extract used in each assay is varied depending on the cell type used or the promoter being studied. For example, 1 μl extract is sufficient in assaying the promoter activity of the RSV LTR (pRSVcat) when introduced into CV-1 or CEF cells (Gorman *et al.*, 1982b). Since the SV40 early promoter does not express well in mouse cells, 30–50 μl extract is used to assay pSV2cat in NIH/3T3 cells. In some experiments, to increase the accuracy of the assay, time points are taken during the 30-min incubation at 37°C. The reaction is stopped with 1 ml cold ethyl acetate, which is also used to extract the chloramphenicol. The organic layer is dried and taken up in 30 μl ethyl acetate, spotted on silica gel TLC plates, and run with chloroform–methanol (95:5, ascending). Following autoradiography of the separated forms of chloramphenicol, spots are quantitated by scintillation counting. Precautions to note when using this assay are: the acetyl-CoA is unstable and therefore should be made up fresh (keep no longer than 2 weeks at $-20°C$) and to prevent any degradation of the [^{14}C]chloramphenicol is stored at $-70°C$.

5. STUDY OF OTHER PROMOTER SEQUENCES

Certain criteria are important when designing a plasmid to test promoter activity. As detailed by Mulligan et al. (1979), the promoter sequence should be subcloned in such a way that the leader sequence of the mRNA remains intact. Though there are examples of translational initiation at an AUG other than the first one following the site of transcriptional initiation (Kozak, 1978; Mulligan and Berg, 1981; Southern and Berg, 1982; Gorman et al., 1982a), it is desirable to try to avoid the inclusion of "extra" AUGs between the leader sequence and the *CAT* start codon (Southern et al., 1981). Therefore, care should be taken in planning the construction of plasmids to test promoter function. To illustrate this, the construction of three plasmids designed to test promoter activity as measured by the CAT assay will be briefly discussed.

A derivative of pSV2cat has been constructed that allows the easy insertion of any putative promoter sequence juxtaposed to the *CAT* gene. In this vector, pSV0cat, the SV40 early promoter has been removed from pSV2cat and reclosure has occurred by reconstruction of a single *Hind*III site (Gorman et al., 1982a). Also available is the vector pSVOcat-Sma, in which a unique *Sma* site has been added next to the *Hind*III site with the use of *Hind*III–*Sma* adaptors (Collaborative Research).

Using the sequence data on the HSV *TK* gene (Wagner et al., 1981), the *TK* promoter was subcloned into a *CAT* vector to give pTKcat (Gorman, Dobson, Khoury, and Howard). A 470-bp *Acc*I–*Hae*II fragment containing the promoter sequence was isolated. Cleavage at the 3′ end of the promoter is approximately 86 bp downstream from the site of mRNA initiation and 15 bp upstream from the *TK* start codon. No AUGs are contained within the leader sequence.

For the construction of pColcat (Ohkubo, Gorman, Howard, and de Crombrugghe), a 1.2-kilobase piece of the chick α2 (I) collagen promoter (Vogeli et al., 1981; Merlino et al., 1981) was subcloned in front of the *CAT* gene. This piece was obtained by a *Hin*fI partial digest followed by a complete *Pvu*II digest. In pColcat, 110 bp of the leader sequence are intact. However, an AUG remains in place preceding the *CAT* start codon. Experiments are in progress to test the effect this may have on *CAT* expression driven by the collagen promoter (McKeon *et al.,* unpublished results).

From the RSV LTR, a 524-bp fragment (*Pvu*II–*Bst*NI) was subcloned to yield pRSVcat (Gorman *et al.,* 1982b). As described above, 30 bp of the leader sequence from the RSV transcriptional unit (Yamamoto *et al.,* 1980) are included. In all the constructs described above, *Hind*III linkers (Collaborative Research) were added to these fragments as described by Gorman *et al.* (1982a).

These three plasmids, pTKcat, pColcat, and pRSVcat, were transfected into CV-1 cells, and the levels of CAT activity were compared to the activity measured from pSV2cat (SV40 early promoter) (Fig. 2). As shown, the HSV TK promoter is relatively weak while the RSV LTR functions as a remarkably strong promoter. More detailed studies on the expression of the RSV LTR

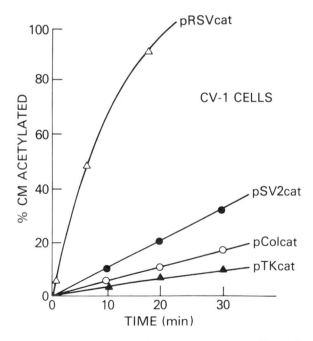

FIGURE 2. Comparison of CAT activity from various promoters. Time points were taken over a 30-min period. Assays were run by the TLC method described and quantitated by scintillation counting. Comparison of the RSV LTR [pRSVcat (△)], the SV40 early promoter [pSV2cat (●)], the chick α2 (I) collagen promoter [pColcat (○)], and the HSV TK promoter [pTKcat (▲)] were made in CV-1 monkey kidney cells.

and the collagen promoter have been published elsewhere (Gorman *et al.*, 1982b).

6. STUDY OF ENHANCING ELEMENTS

The CAT system has also been used to quantitate the role that enhancer sequences have in the level of transcription. Deletions were made in the early region of SV40, either one or both of the 72-bp repeats being removed. The level of CAT activity produced by the plasmid containing only one repeat [p(sph)cat] is indistinguishable from the parent pSV2cat (Fig. 3) (Gorman *et al.*, 1982a). This is consistent with findings of others (Gruss *et al.*, 1981; Moreau *et al.*, 1981). The sensitivity of this system has enabled the detection of transcription from the SV40 early promoter lacking both the enhancer repeats in pSV1cat (Gorman *et al.*, 1982a) and $pA_{10}cat_2$ (Laimons *et al.*, 1982) (Fig. 3). Other attempts to measure transcription from the SV40 early promoter when both repeats have been deleted have been negative (Gruss *et*

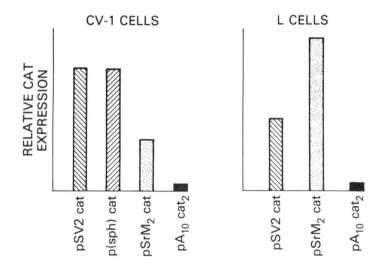

FIGURE 3. Effects of various enhancing sequences on CAT production. Relative levels of CAT produced by pSV2cat and modified vectors are shown in both CV-1 monkey kidney cells and mouse L cells. Plasmid p(sph)cat retains one copy of the 72-bp repeat of SV40. Plasmid $pA_{10}cat_2$ has both the SV40 repeats removed. Plasmid $pSrM_2cat$ contains the tandem 72-bp repeats from MSV.

al., 1981; Moreau *et al.*, 1981). The vector $pA_{10}cat_2$ contains a single *Bg*/II site at the deletion site of the 72-bp repeats (Laimons *et al.*, 1982). Using this vector, other potential enhancer sequences can be tested. The enhancer repeats from the MSV LTR (Levinson *et al.*, 1982) have replaced the SV40 repeats to give $pSrM_2cat$ (Laimons *et al.*, 1982). The vectors pSV2cat and $pSrM_2cat$ differ only in the enhancer sequence; both retain the SV40 early "TATA" box and initiation site for transcription. These two plasmids were used to compare the enhancing effect of the repeats from SV40 and MSV using the same "TATA" sequence. There is a relative difference in the level of CAT activity from these two plasmids depending on what cell type is used (Fig. 3) (Laimons *et al.*, 1982).

7. CONCLUSION

We have described an assay system that uses the expression of a bacterial gene, chloramphenicol acetyltransferase, to study eukaryotic promoter activity. The absence of this enzyme from any eukaryotic cells greatly adds to the sensitivity and use of the CAT assay. This approach has facilitated the detection of gene products from weakly expressing genes. Other assays have been described (Mulligan and Berg, 1980; Schumperli *et al.*, 1982; Summers and Summers, 1977); however, with these systems, one must distinguish between the endogenous and exogenous enzyme activity.

Using viral promoters that have been shown to be quite strong, RNA levels correlate with levels of CAT activity (Gorman *et al.*, 1982b; Laimons *et al.*, unpublished results), though there may be some difference in translational efficiency between promoters. In studying less active cellular promoters, isolation and quantitation of RNA can be difficult. This has been true with the chick $\alpha2$ (I) collagen promoter (Ohkubo *et al.*, unpublished results) even though the CAT levels are easily quantitated (see Fig. 2).

In addition to the ease and sensitivity of the CAT assay, we have used fluorescent anti-CAT to monitor transfection efficiencies (Gorman *et al.*, 1982b). By this means, we have determined that the various plasmids studied are not differentially taken up. The transfection efficiency can vary from experiment to experiment, and this can affect the absolute amount of CAT activity, though we have found that the relative amounts remain constant when comparing plasmids. There are precautions one can take to keep variation at a minimum. As stated above, cells must be in log phase during the uptake of DNA. Using a constant number of cells helps to reduce fluctuations between experiments. We find that twice-banded DNA forms very fine precipitates, and this seems to increase transfection efficency. Our experience in these studies is that the CAT system provides a rapid and accurate method to compare different promoters and enhancing sequences, to measure the apparent strength of one promoter in a number of cell types, to compare transfection protocols, and to evaluate other parameters that affect exogenous gene expression.

REFERENCES

Benoist, C., and Chambon, P., 1980, Deletions covering the putative promoter region of early mRNAs of simian virus 40 do not abolish T-antigen expression, *Proc. Natl. Acad. Sci. U.S.A.* **77**:3865–3869.

Benoist, C., and Chambon, P., 1981, *In vivo* sequence requirements of the SV40 early promoter region, *Nature (London)* **290**:304–310.

Cohen, J., Eccleshall, T., Needleman, R., Federoff, H., Buchferer, B., and Marmur, J., 1980, Functional expression in yeast of the *Escherichia coli* plasmid gene coding for chloramphenicol acetyltransferase, *Proc. Natl. Acad. Sci. U.S.A.* **77**:1078–1082.

De Villiers, J., and Schaffner, W., 1981, A small segment of polyoma virus DNA enhances the expression of a cloned β-globin gene over a distance of 1400 base pairs, *Nucleic Acid Res.* **9**:6251–6264.

Gorman, C., Moffat, L., and Howard, B., 1982a, Recombinant genomes which express chloramphenicol acetyltransferase in mammalian cells, *Mol. Cell. Biol.* **2**:1044–1051.

Gorman C., Merlino, G., Willingham, M., Pastan, I., and Howard, B., 1982b, The Rous sarcoma virus long terminal repeat is a strong promoter when introduced into a variety of eukaryotic cells by DNA mediated transfection, *Proc. Natl. Acad. Sci. U.S.A.* **77**:6777–6781.

Graham, F., and van der Eb, A., 1978, A new technique for the assay of infectivity of human adenovirus 5 DNA, *Virology* **52**:456–457.

Gruss, P., Dhar, R., and Khoury, G., 1981, Simian virus 40 tandem repeated sequences as an element of the early promoter, *Proc. Natl. Acad. Sci. U.S.A.* **78**:943–947.

Kozak, M., 1978, How do eukaryotic ribosomes select initiation regions in messenger RNA?, *Cell* **15**:1109–1123.

Laimons, L., Khoury, G., Gorman, C., Howard, B., and Gruss, P., 1982, Host specific activation of transcription by tandem repeats from SV40 and Moloney murine sarcoma virus, *Proc. Natl. Acad. Sci. U.S.A.* **79**:6453–6457.

Levinson, B., Khoury, G., Vande Woude, G., and Gruss, P., 1982, Activation of SV40 genome by 72-base pair tandem repeats of Moloney sarcoma virus, *Nature (London)* **295**:568–572.

Manley, J., Fire, A., Cano, A., Sharp, P., and Gefter, M., 1980, DNA-dependent transcription of adenovirus genes in a soluble whole-cell extract, *Proc. Natl. Acad. Sci. U.S.A.* **77**:3855–3859.

Mellon, P., Parker, V., Gluzman, Y., and Maniatis, T., 1981, Identification of DNA sequences required for transcription of the human α1 globin gene in a new SV40 host–vector system, *Cell* **27**:279–288.

Merlino, G., Vogeli, G., Yamamoto, T., de Crombrugghe, B., and Pastan, I., 1981, Accurate *in vitro* transcriptional initiation of the chick α2 (I) collagen gene, *J. Biol. Chem.* **256**:11251–11258.

Moreau, P., Hen, R., Wasylyk, B., Everett, R., Gaub, M., and Chambon, P., 1981, The SV40 72 base pair repeat has a striking effect on gene expression both in SV40 and other chimeric recombinants, *Nucleic Acids Res.* **9**:6047–6068.

Mulligan, R., and Berg, P., 1980, Expression of a bacterial gene in mammalian cells, *Science* **209**:1422–1427.

Mulligan, R., and Berg, P., 1981, Factors governing the expression of bacterial genome, *Mol. Cell. Biol.* **1**:449–459.

Mulligan, R., Howard, B., and Berg, P., 1979, Synthesis of rabbit β-globin in cultured monkey cells following infection with a SV40 β-globin recombinant gene in mammalian cells, *Nature (London)* **277**:108–114.

Parker, B., and Stark, G., 1979, Regulation of simian virus 40 transcription: Sensitive analysis of the RNA species present early in infections by virus or viral DNA, *J. Virol.* **31**:360–369.

Radloff, R., Bauer, W., and Vinograd, J., 1967, A dye buoyant-density method for the detection and isolation of closed circular duplex DNA: The closed circular DNA in HeLa cells, *Proc. Natl. Acad. Sci. U.S.A.* **57**:1514–1521.

Robinson, L., Seligohn, R., and Lerner, S., 1978, Simplified radioenzymatic assay for chloramphenicol, *Antimicrob. Agents Chemother.* **13**:25–29.

Schumperli, D., Howard, B., and Rosenberg, M., 1982, Efficient expression of *Escherichia coli* galactokinase gene in mammalian cells, *Proc. Natl. Acad. Sci. U.S.A.* **79**:257.

Scott, J., 1973, Phage P1 cryptic: Location and regulation of prophage genes, *Virology* **53**:327–336.

Shaw, W., 1967, The enzymatic acetylation of chloramphenicol by extracts of R factor-resistant *Escherichia coli*, *J. Biol. Chem.* **242**:687–693.

Shaw, W., 1975, Chloramphenicol acetyltransferase from resistant bacteria, *Methods Enzymol.* **53**:737–754.

Shaw, W., and Brodsky, R., 1978, Characterization of chloramphenicol acetyltransferase from chloramphenicol resistant *Staphylococcus aureus*, *J. Bacteriol.* **95**:28–36.

Smith, A., and Smith, P., 1978, Improved enzymatic assay of chloramphenicol, *Clin. Chem.* **24**:1452–1457.

Southern, P., and Berg, P., 1982, Transformation of mammalian cells to antibiotic resistance with a bacterial gene under control of the SV40 early region promoter, *J. Mol. Appl. Genet.* **1**:327–341.

Southern, P., Howard, B., and Berg, P., 1981, Construction and characterization of SV40 recombinants with β-globin cDNA substitutions in their early region, *J. Mol. Appl. Genet.* **1**:177–190.

Summers, W., and Summers, W., 1977, 125-I deoxycytidine used in a rapid, sensitive, and specific assay for herpes simplex virus type 1 thymidine kinase, *J. Virol.* **24**:314–318.

Vogeli, G., Ohkubo, H., Sobel, M., Yamada, Y., Pastan, I., and de Crombrugghe, B., 1981, Structure of the promoter of chicken α2 type I collagen gene, *Proc. Natl. Acad. Sci. U.S.A.* **78**:5334–5338.

Wagner, M., Sharp, J., and Summers, W., 1981, Nucleotide sequence of the thymidine kinase gene of herpes simplex virus type 1, *Proc. Natl. Acad. Sci. U.S.A.* **78**:1443–1445.

Weil, P., Luse, D., Segall, J., and Roeder, R., 1979, Selective and accurate initiation of transcription at the Ad 2 major late promoter on a soluble system dependent on purified RNA polymerase II and DNA, *Cell* **18**:469–484.

Yamamoto, T., de Crombrugghe, B., and Pastan, I., 1980, Identification of a functional promoter of Rous sarcoma virus, *Cell* **22**:787–797.

Ti Plasmids as Gene Vectors for Plants

JEFF SCHELL, MARC VAN MONTAGU,
MARCELLE HOLSTERS,
PATRICIA ZAMBRYSKI, HENK JOOS,
LUIS HERRERA-ESTRELLA, ANN DEPICKER,
JEAN-PIERRE HERNALSTEENS,
HENRI DE GREVE, LOTHAR WILLMITZER,
and JO SCHRÖDER

1. INTRODUCTION

The formation of so-called "crown gall" tumors on dicotyledonous plants is the direct result of the introduction into the nuclear genome of plant cells of a set of genes that regulate cell and organ development. In other words, in nature, a mechanism exists that not only efficiently introduces foreign genes into the plant nucleus, but also contains a set of genes that regulate plant-cell development and differentiation. As a result of this gene transfer, crown gall cells, unlike untransformed plant tissues, can be cultured under axenic con-

JEFF SCHELL ● Laboratorium voor Genetica, Rijksuniversiteit Gent, B-9000 Gent, Belgium; Max-Planck-Institut für Züchtungsforschung, D-5000 Cologne 30, Federal Republic of Germany. MARC VAN MONTAGU ● Laboratorium voor Genetica, Rijksuniversiteit Gent, B-9000 Gent, Belgium; Laboratorium voor Genetische Virologie, Vrije Universiteit Brussel, B-1640 Sint-Genesius-Rode, Belgium. MARCELLE HOLSTERS, PATRICIA ZAMBRYSKI, HENK JOOS, LUIS HERRERA-ESTRELLA, and ANN DEPICKER ● Laboratorium voor Genetica, Rijksuniversiteit Gent, B-9000 Gent, Belgium. JEAN-PIERRE HERNAL-STEENS and HENRI DE GREVE ● Laboratorium voor Genetische Virologie, Vrije Universiteit Brussel, B-1640 Sint-Genesius-Rode, Belgium. LOTHAR WILLMITZER and JO SCHRÖDER ● Max-Planck-Institut für Züchtungsforschung, D-5000 Cologne 30, Federal Republic of Germany.

ditions on synthetic media in the absence of growth hormones, i.e., cytokinins and auxins.

This process is carried out by the so-called Ti plasmids of *Agrobacterium*. Most strains of the *Agrobacterium* genus, both pathogenic and nonpathogenic, contain one or more large plasmids (Zaenen *et al.*, 1974; Currier and Nester, 1976; Merlo and Nester, 1977), many of which have remained uncharacterized. The different types of Ti plasmids that are responsible for the pathogenic properties of *Agrobacterium* all have a molecular weight in the range of $120–160 \times 10^6$ daltons. A subgroup of Ti plasmids inducing hairy root tumors are often referred to as Ri plasmids. The tumor-inducing plasmids are most easily identified by transferring to nonvirulent strains of bacteria (Van Larebeke *et al.*, 1974; Watson *et al.*, 1975). The transfer of virulence is always correlated with the transfer of a Ti plasmid (Kerr, 1969, 1971; Bomhoff *et al.*, 1976; Hooykaas *et al.*, 1977; Van Larebeke *et al.*, 1977; Genetello *et al.*, 1977; Kerr *et al.*, 1977; Holsters *et al.*, 1978). The tumor cells also produce low-molecular-weight compounds, called opines, not found in untransformed plant tissues. The type of opine produced defines crown galls as octopine-, nopaline-, or agropine-type tumors (Guyon *et al.*, 1980). Transfer experiments have demonstrated that Ti plasmids are responsible for most of the typical properties of agrobacteria: (1) crown gall tumor induction, (2) specificity of opine synthesis in transformed plant cells, (3) catabolism of specific opines, (4) agrocin sensitivity, (5) conjugative transfer of Ti plasmids, and (6) catabolism of arginine and ornithine (Zaenen *et al.*, 1974; Van Larebeke *et al.*, 1974, 1975; Watson *et al.*, 1975; Bomhoff *et al.*, 1976; Genetello *et al.*, 1977; Kerr *et al.*, 1977; Guyon *et al.*, 1980; Schell, 1975; Engler *et al.*, 1975; Petit *et al.*, 1978a,b; Firmin and Fenwick, 1978; Klabwijk *et al.*, 1978; Ellis *et al.*, 1979). The crown gall tumors contain a DNA segment (called T-DNA) derived from Ti plasmids that is homologous and colinear with a defined fragment of the corresponding Ti plasmid present in the tumor-inducing bacterium (Chilton *et al.*, 1977; Schell *et al.*, 1979; Lemmers *et al.*, 1980; Thomashow *et al.*, 1980a; De Beuckeleer *et al.*, 1981). The T-DNA is covalently linked to plant DNA (Zambryski *et al.*, 1980; Yadav *et al.*, 1980; Thomashow *et al.*, 1980b) in the nucleus of the plant cell (Chilton *et al.*, 1980; Willmitzer *et al.*, 1980). The T-DNA is transcribed in the transformed plant cell, and T-DNA-encoded proteins, such as the octopine-synthesizing enzyme, lysopine dehydrogenase (LpDH), have been found in the several octopine crown gall lines (J. Schröder *et al.*, 1981a). This has led to the realization that Ti plasmids are a natural gene vector for plant cells, evolved by and for the benefit of the bacteria that harbor Ti plasmids (Schell *et al.*, 1979). All the genetic information for the synthesis of opines in transformed plant cells and for their catabolism by free-living agrobacteria is carried by Ti plasmids. In this way, free-living agrobacteria can utilize as sources of carbon and nitrogen the opines produced by the tumors they have incited.

   In this chapter, we want to concentrate on those aspects of the structure and the properties of Ti plasmids that are important for their use as experimental gene vectors.

2. INTEGRATION OF A SEGMENT OF THE Ti PLASMID (T-REGION) INTO PLANT NUCLEAR DNA

   The T-region is defined as that segment of the Ti plasmids that is homologous to sequences present in crown gall cells. The sequences that are transferred from the Ti plasmid to the plant and determine tumorous growth have been called T-DNA. The T-regions of octopine and nopaline Ti plasmids have been studied in great detail both physically and functionally. The T-regions, roughly 23 kilobases (kb) in size, are only a portion of the entire plasmids (Lemmers *et al.*, 1980; Thomashow *et al.*, 1980a; De Beuckeleer *et al.*, 1981; Engler *et al.*, 1981). Southern blotting and cross-hybridization of restriction-endonuclease digests of the two types of plasmids as well as electron-microscopic heteroduplex analyses have revealed that 8–9 kb of the T-DNA regions are conserved and common to both octopine and nopaline types of plasmids (Engler *et al.*, 1981; Chilton *et al.*, 1978; Depicker *et al.*, 1978). DNA sequence data confirm that these "common" or "core" segments of the T-regions are about 90% homologous with one another (unpublished results). It was postulated early on (Schell and Van Montagu, 1977; Chilton *et al.*, 1978) that this common core might contain genes essential for tumor formation and maintenance. Recent data on the expression of the T-DNA and the study of the effects of insertion and deletion mutations in the common core of the T-region have verified and extended this hypothesis (see following sections). Attempts to reveal homology between the T-DNA region and plant DNA have failed thus far.

   Detailed analysis of some nopaline lines (Lemmers *et al.*, 1980; Zambryski *et al.*, 1980, 1982; Yadav *et al.*, 1980) suggests that the mechanism of T-DNA integration is rather precise, since the same continuous segment of the Ti plasmid is always present. Some lines appear to contain a single T-DNA copy, whereas in others the T-DNA occurs in multiple copies that are organized in a tandem array.

Several octopine tumor lines have been studied, and the data suggest that the octopine T-DNA is more variable (Thomashow *et al.*, 1980a,b; De Beuckeleer *et al.*, 1981). A left T-DNA region (TL) containing the "common" or "core" sequence is always present; this region is usually 12 kb in size, but one *Petunia* tumor line is shortened at the right end of TL by about 4 kb (De Beuckeleer *et al.*, 1981). In addition, there is often a right T-DNA region (TR) that contains sequences that are adjacent but not contiguous in the octopine Ti plasmid. In some tumor lines, TR is amplified whereas TL is not (Thomashow *et al.*, 1980a; Merlo *et al.*, 1980). Recent observations demonstrate that TL can also be part of a tandem array (Holsters *et al.*, 1983). It is not known whether the integration is the result of plant- or Ti-plasmid-specific functions, but it is likely that both are involved. Figure 1 illustrates the findings regarding the T-DNA borders in tumors induced with nopaline Ti plasmids. The 25-base-pair (bp) direct repeat thought to determine the T-DNA borders in octopine and nopaline tumors is illustrated in Fig. 2.

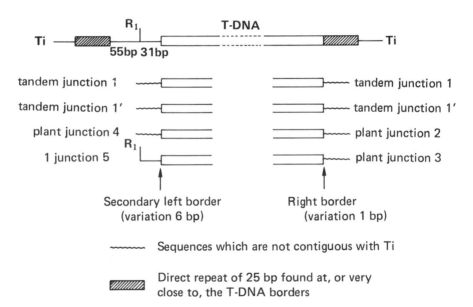

FIGURE 1. Summary of nopaline T-DNA borders compared to the nopaline Ti plasmid. The borders of nopaline T-DNAs are diagrammed in comparison to the border regions of the nopaline Ti plasmid; the exact nucleotide sequences of these DNAs have been published (Zambryski *et al.*, 1982). The origin of clones 1, 1′, 2, 3, 4, and 5 is described in the text. One *Eco*RI restriction-endonuclease site that is close to the left T-DNA border is shown. In addition, the relative position of the direct repeat of 25 bp shown in Fig. 2 is indicated.

```
GCTGG   TGGCAGGATATATTG    TG   GTGTAAAAC   AAATT   Nopaline L
GTGTT   TG(A)CAGGATATATTG  GC   G(G)GTAAAAC CTAAG   Nopaline R
AGCGG   (G)GGCAGGATATATTC  AA   TTGTAAAT    GGCTT   Octopine L (TL)
CTGAC   TGGCAGGATATAT(A)C  CG   TTGTAA(T)T  TGAGC   Octopine R (TL)
```

FIGURE 2. Summary of the 25-bp direct repeat found at the T-DNA borders of the nopaline and octopine Ti plasmids. A direct repeat of 25 bp that is found at or very close to the T-DNA borders in nopaline and octopine Ti plasmids is shown. In the nopaline Ti plasmid (pTiC58 or pTiT37), this repeat occurs exactly at the right T-DNA border and starts 111 bp to the left of the secondary left border. In the octopine Ti plasmid (pTiAch5 or pTiA6S3), this repeat occurs very close to and just outside the right T-DNA border; two T-DNA junctions analyzed so far vary within 6 bp from this sequence (Holsters et al., 1983). On the left, in one tumor line this sequence occurs exactly at and within the T-DNA border (Simpson et al., 1982). In two other tumor lines, the left T-DNA border occurs either within or 59 bp internal to this sequence (Holsters et al., 1983).

In view of the observed involvement of the "ends" of the T-region in the integration of T-DNA, it was expected that any DNA segment inserted between these "ends" would be cotransferred, provided no function essential for T-DNA transfer and stable maintenance was inactivated by the insertion. The genetic analysis of the T-region by transposon insertion provided Ti plasmid mutants to test this hypothesis. A Tn7 insertion in the nopaline synthase locus produced a Ti plasmid able to initiate T-DNA transfer and tumor formation (De Beuckeleer *et al.*, 1978; Van Montagu and Schell, 1979; Hernalsteens *et al.*, 1980).

Analysis of the DNA extracted from these tumors showed that the T-region containing the Tn7 sequence had been transformed as a single 38×10^3-bp segment without any major rearrangements. Several different DNA sequences have since been introduced into different parts of the T-region of octopine and nopaline Ti plasmids (Leemans *et al.*, 1981, 1982; Joos *et al.*, 1983). The preliminary observations with these mutant Ti plasmids fully confirm our initial expectations: as a rule, DNA sequences inserted between the "ends" of the T-region are cotransferred with, and become a stable part of, the T-DNA of the plant tumor cells transformed with these mutant Ti plasmids. If the experimental insert were to inactivate a function essential for the transfer of the T-region, or for the integration of the T-DNA, such a mutant Ti plasmid would not be able to transform plant cells. With one possible exception, no such inserts have as yet been characterized, indicating that T-DNA transfer and integration are probably coded for by genes outside the T-region or by a combination of functions with different genetic localization (Leemans *et al.*, 1982; Joos *et al.*, 1983).

These observations have therefore firmly established that the Ti plasmids can be used as experimental gene vectors and that large DNA sequences (up to 50 kb) can be transferred stably to the nucleus of plant cells as a single DNA segment.

3. MODIFICATION OF Ti PLASMIDS WITH THE PURPOSE OF ELIMINATING THE TUMOR-CONTROLLING PROPERTIES

A double approach was used to elucidate the mechanism of tumor formation resulting from the integration of T-DNA in the plant-cell nucleus. It was first shown that the T-DNA consists of a number of well-defined transcriptional units, which were carefully located on the physical map of the corresponding T-region. Subsequently, insertions and deletion substitutions were introduced at specific sites of the T-region to produce Ti plasmids carrying mutant T-regions such that one or another or a combination of several transcripts could not be formed in plant cells carrying such mutant T-DNAs.

RNA transcripts homologous to T-DNA have been shown to be present in all crown gall tissues studied thus far (Drummond *et al.*, 1977; Gurley *et al.*, 1979; Willmitzer *et al.*, 1981a; Gelvin *et al.*, 1981). The number, sizes, and location of the transcribed T-DNA segments were studied in both octopine and nopaline tumors (Willmitzer *et al.*, 1982a, 1983). Tumor-specific RNAs were detected and mapped by hybridization of ^{32}P-labeled Ti plasmid fragments to RNA that had been separated on agarose gels and then transferred to diazobenzoyloxymethyl paper. The results show that the octopine tumors contain a total of 12 distinct transcripts (8 different transcripts are derived from TL-DNA and 4 from TR-DNA), whereas nopaline tumors were found to contain at least 13 different transcripts.

These transcripts differ in their relative abundance and in their sizes. They all bind to oligodeoxythymidylic acid–cellulose, indicating that they are polyadenylated. Thus, the T-DNA transferred from a prokaryotic organism provides specific polyadenylic acid [poly(A)] addition sites. The direction of transcription was determined, and the locations of the approximate 5' and 3' ends were mapped on the T-DNA.

All RNAs mapped within the T-DNA sequence. This, and the observation that transcription is inhibited by low concentrations of α-amanitin (Willmitzer *et al.*, 1981b), suggest that each transcript is determined by a specific promoter site on the T-DNA recognized by plant RNA polymerase II. Since not all transcripts were synthesized from the same strand of DNA, the simplest model for transcription would be that there is one promoter site per group of transcripts. If so, one would expect that the deletion of a 5'-proximal gene of a group would also lead to the disappearance of the transcripts from the 3'-distal genes. However, analysis of cell lines containing the T-DNA of Ti plasmid mutants indicated that genes could not be inactivated by mutations lying far from the coding region (Leemans *et al.*, 1982; Joos *et al.*, 1983). The results available so far are consistent with the assumption that each gene on the T-DNA has its own signals for transcription in the eukaryotic plant cells. Six transcripts were found to be derived from the "common" or "core" segment of the T-region. These transcripts were found to be identical in nopaline and octopine tumors (Willmitzer *et al.*, 1982a, 1983).

To determine whether these messenger RNAs (mRNAs) are translated into proteins, a hybridization selection procedure was developed that was sufficiently sensitive and specific to detect mRNAs that represent about 0.0001% of the total mRNA activity in the plant cell (the concentration of total T-DNA-specific RNA in the octopine tumor line A6-S1 is between 0.0005 and 0.001%). This procedure was used to enrich for T-DNA-derived mRNAs by hybridization to Ti plasmid fragments covalently bound to microcrystalline cellulose; the hybridized RNAs were eluted and translated *in vitro* in a cell-free system prepared from wheat germ.

The results obtained with this approach (G. Schröder and J. Schröder, 1982) showed that tumor cells contain at least three T-DNA-derived mRNAs that can be translated *in vitro* into distinct proteins. The protein encoded at the right end of the TL-DNA (molecular weight 39,000 daltons) was of particular interest, since previous genetic analysis indicates that this region is responsible for octopine synthesis (Koekman *et al.*, 1979; Garfinkel *et al.*, 1981; De Greve *et al.*, 1981).

The *in vitro*-synthesized protein was shown to be identical in size with the octopine-synthesizing enzyme in octopine tumors. Immunological studies showed that this protein was recognized by antiserum against the tumor-specific synthase (J. Schröder *et al.*, 1981a). These results demonstrate that the structural gene for the octopine-synthesizing enzyme is on the Ti plasmid. So far, this is the only protein product of the T-DNA with known enzymatic properties; the possible functions of two smaller T-DNA-derived proteins are not known. The region coding for the octopine-synthesizing enzyme has recently been sequenced (De Greve *et al.*, 1982a). The 5' end of the octopine synthase mRNA was accurately mapped by sequencing a T-region DNA fragment that hybridizes to this mRNA and thus protects it from degradation by the single-strand-specific S1 nuclease. The promoter sequence thus identified is more eukaryotic than prokaryotic in its recognition signals, and no introns interrupt the open-reading frame that starts at the first AUG codon following the 5' start of the transcript. Similar work leading to essentially the same conclusions has also been performed for the nopaline synthase gene (Depicker *et al.*, 1982). It is important to note that the promoter sequences for the octopine synthase appear to escape possible control mechanisms, since they remain active in all tissues of plants regenerated from tobacco cells transformed with mutant Ti plasmids (De Greve *et al.*, 1982b).

The 3'-polyadenylated terminus of the transcript was also analyzed, and a polyadenylation signal 5'-AAUAA-3' was found about 10 nucleotides from the start of the poly(A) sequence. This appears to be a general feature of eukaryotic mRNAs, since it has been observed in a number of animal mRNAs (Fitzgerald and Shenk, 1981). To some extent, therefore, these opine-synthesizing genes seem designed to function in eukaryotic cells rather than in prokaryotic cells.

However, this is not necessarily true for all genes of the T-region, since transcripts were also detected in agrobacteria (Gelvin *et al.*, 1981, and unpublished data). For this reason, it was interesting to determine whether all the T-DNA-derived mRNAs isolated from plant cells shared properties with typical eukaryotic mRNAs. The fact that translation of each of the three mRNAs analyzed *in vitro* was inhibited by the cap analogue pm^7G suggests that they contain a cap structure at the 5' end. This would be typical for eukaryotic mRNA, since caps have not been described in prokaryotic RNA.

As noted above, each of the mRNAs of the three *in vitro*-synthesized proteins represents about 0.0001% of the total mRNA activity in polyribosomal RNA, and this appears to be the detection limit at present for translatable RNA. Some of the other transcripts detected by hybridization experiments are present at even lower concentrations. Assuming that they possess mRNA activity, this is likely to be the reason that it has not yet been possible to identify the corresponding proteins by *in vitro* translation.

A different approach has been developed to search for coding regions on the T-DNA and their protein products. Fragments from the T-region were cloned into *Escherichia coli* plasmids and analyzed for gene expression in *E. coli* minicells (J. Schröder *et al.*, 1981b). There are at least four different coding regions within the TL-DNA that can be expressed from promoters that are active in prokaryotic cells and translated into proteins in minicells. The four regions expressed in *E. coli* correlate with four regions transcribed into RNA in plant cells. The plant transcripts are larger than the proteins in *E. coli,* and the regions expressed in minicells appear to lie within the regions transcribed in plant cells. One can therefore speculate that plant cells and *E. coli,* at least partly, express the same coding regions.

Specific mutations were introduced in the T-DNA regions of octopine and nopaline Ti plasmids to produce transformed plant cells in which one or more T-DNA-derived transcripts would not be expressed. By observing the phenotypes of the plant cells harboring such partially inactivated T-DNAs, it was possible to assign functions to most of the different transcripts (Leemans *et al.*, 1982; Joos *et al.*, 1983). It was found that none of the T-DNA transcripts was essential for the transfer and stable maintenance of T-DNA segments in the plant genome. Essentially, two different functions were found to be determined by T-DNA transcripts:

1. Transcripts coding for opine synthase. Octopine tumors contain either one or two such genes. One is located on the right end of TL and codes for

octopine synthase, the other is located at the right end of TR and codes for agropine and mannopine synthase (J. Velten, personal communication). Tumors that contain both TL and TR therefore produce both octopine and agropine. Nopaline tumors also contain at least two transcripts coding for different opines (Joos *et al.*, 1983). One is located at the right end of the T-DNA and codes for nopaline synthase, whereas the other is located in the left part of the T-DNA and codes for agrocinopine.

2. Transcripts (probably after translation into proteins) that are directly or indirectly responsible for tumorous growth. These transcripts are found to be derived from the "common" or "core" region of the T-DNA (Fig. 3). In total, six different well-defined transcripts were found to be derived from this "common" region. Remarkably, all T-DNA functions affecting the tumor phenotype were located in this "common" region of the T-DNA. Several of these transcripts act by suppressing plant organ development. It was observed that shoot and root formation are suppressed independently and by different transcripts. Two transcripts (1 and 2) were identified that specifically prevent shoot formation. The effect of these T-DNA gene products is in many ways analogous to that of auxinlike plant growth hormones, since the effect of these genes is similar to that observed for calli from normal plant cells with artificially increaed auxin level. Another transcript (transcript 4) was found to specifically prevent root formation, and the effect of this T-DNA gene can therefore be compared to the effects observed when normal plant cells are grown in the presence of high concentrations of cytokinins.

That both the shoot and root inhibition resulting from the activity of these genes may be due to the fact that they directly or indirectly determine the formation of auxin- and cytokininlike growth hormones (Skoog and Miller, 1957; Ooms *et al.*, 1981) is further substantiated by our observation that these genes respectively inhibit shoot or root formation in both T-DNA-containing and T-DNA-negative (normal) cells, provided both types of cells grow as one mixed tissue.

This interpretation of the possible function of transcripts 1 and 2 (auxinlike) and of transcript 4 (cytokininlike) is consistent with recent measurements of endogenous levels of auxin and cytokinin in teratoma and unorganized crown gall tissue (Amasino and Miller, 1982).

Even more convincing are the observations (Morris *et al.*, 1982) that the cytokinin/auxin ratio in wild-type tumors is 0.22, whereas it is 0.02 in rooting tumors induced by T-region mutants in gene 4 and 14.4 in shooting tumors induced by Ti mutants in the region of gene 1 or 2. However, a word of

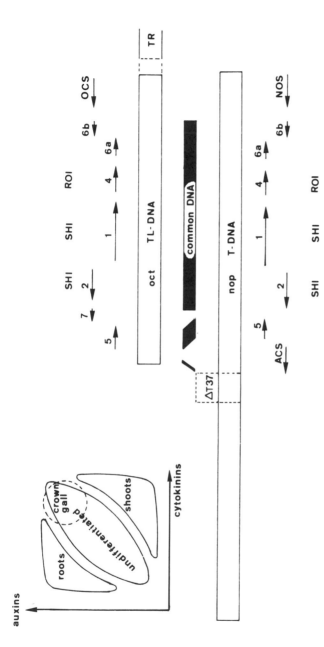

FIGURE 3. Octopine and nopaline T-DNA maps with their localized transcripts. Their homology (Engler et al., 1981) is indicated as a bar between both T-DNAs with the gene they specify. The eight transcripts are mapped on the octopine T-DNA according to Willmitzer et al. (1982a). Only the six nopaline T-DNA transcripts homologous with the octopine T-DNA are indicated; the other nopaline T-DNA transcripts will be presented elsewhere (Willmitzer et al., 1983). SHI (shoot inhibition), ROI (root inhibition), OCS (octopine synthase), ACS (agrocinopine synthase), and NOS (nopaline synthase) are phenotypes corresponding to the mapped loci on the T-DNA. The auxin–cytokinin diagram is a schematic representation of how the hormone levels can influence the callus morphology (Skoog and Miller, 1957). Variations of the auxin/cytokinin ratio cause the callus to give rise to buds, shoots, or roots. A similar control could be exerted by the T-DNA in a direct or indirect way. Inactivation of T-DNA transcripts can cause a different crown gall morphology, possibly as an alteration in the auxin/cytokinin ratio.

caution should be added. Indeed, whereas all these observations are consistent with the idea that the products of genes 1, 2, and 4 directly determine the auxin–cytokinin levels in the transformed cells and that these hormone levels in turn would be responsible for the observed tumor phenotypes, they do not prove this point. It is, for instance, still conceivable that the products of genes 1, 2, and 4 could act directly at the level of gene regulation and that the alterations in growth hormone levels would be the consequence, rather than the cause, of the observed tumor morphology. It is essential to isolate these gene products to determine in detail their mechanism of action. In addition to this hormonelike activity, the T-DNA codes for at least three other transcripts, 5, 6, and 6a. The combination of two of these transcripts, 6 and 6a, with the cytokininlike activity of transcript 4 was shown to be sufficient to suppress development of transformed cells and to allow their hormone-independent growth (Leemans *et al.*, 1982). Another transcript, 5, was found to inhibit the organization of transformed cells into leaf bud structures. Elimination of this transcript, along with the shoot-inhibiting auxinlike genes (genes 1 and 2), resulted in transformed cells organizing themselves as teratomas (Leemans *et al.*, 1982; Joos *et al.*, 1983).

Whereas the hormonelike effect of genes 1, 2, and 4 results in a suppression of regeneration by both non-T-DNA-containing and T-DNA-containing plant cells, the effect of gene 5 seems to be restricted to the plant cells in which this gene is present and active.

Further evidence in favor of the idea that the *"onc"* genes of the T-DNA are responsible for tumor formation primarily because they negatively control (suppress) differentiation of shoots and roots is based on an analysis of spontaneous deletion mutants.

It is observed that untransformed cells, mixed in primary tumors with T-DNA-containing transforming cells, are able to regenerate normal plants provided that genes 1 or 2, or both, of the transformed cells are inactive. The transformed cells themselves are still suppressed for regeneration, but it seems reasonable that if spontaneous mutations were to inactivate the genes responsible for the suppression of the transformed cells, then such cells would also be able to regenerate and form shoots and roots. To recognize such plants derived from cells containing mutated T-DNAs, a large number of shoots from a shooting octopine tumor were screened for the presence of octopine synthase. Most of the shoots were negative, but some of the proliferating shoots were positive. Several of these shoots were grown further on growth-hormone-free media and found to develop roots and, later, to grow in fully normal, flowering plants. Each part of these plants—leaves, stem, and roots—was found to contain octopine synthase activity, and polysomal RNA was found to contain T-DNA transcripts homologous to the opine synthesis locus. No transcripts of the common segment of the T-region were observed. One of these plants, rGV1, was studied in great detail (De Greve *et al.,* 1982b), and its T-DNA was isolated from the plant DNA by molecular cloning in a λ phage vector. The T-DNA was found to have undergone a large deletion, removing all but the rightmost part of the T-DNA that codes for the octopine synthase gene. This explains the transcription data of LpDH-positive plants. On the basis of this tissue line, it would appear that for fully normal plants to be formed by T-DNA-containing cells, it is essential that genes 1, 2, and 4, and possibly genes 5, 6a, and 6b, be inactivated.

We are uncertain about genes 6a and 6b because no important phenotypic change has thus far been correlated with their inactivation (Joos *et al.,* 1983).

Apparently, a different conclusion could result from recent observations by Barton *et al.* (1983). These authors found that they could regenerate normally organized plants from nopaline-positive roots obtained by cellular cloning of a rooting tumor. This rooting tumor was induced by a nopaline T-region mutant in which only gene 4 is inactive. These "normal" nopaline-positive plants were analyzed by DNA–DNA hybridizations and were found to contain all the mutated T-region. In fact, the T-DNA appeared to be amplified to 20 copies per genome in these plants.

Since no transcription studies have yet been done with these plants, it is still conceivable that the other genes of the common T-region are inactive in these plants due to a mutation other than a deletion that would not be detected by Southern gel blotting and DNA–DNA hybridizations. Alternatively, the prior inactivation of gene 4 could create the proper conditions for regeneration even in the presence of genes 5, 1, and 2. These observations, together with ours (Leemans *et al.,* 1982) showing that hormone-independent, teratomalike tumorous growths can be obtained by the presence solely of genes 4, 6a, and 6b (we do not yet know whether genes 6a and 6b play any significant role here), indicate that the product of gene 4 is probably the most important for the formation and the maintenance of the tumorous state in Ti-plasmid-transformed plant cells.

Support for this hypothesis comes from the observation (J. Tempé, personal communication) that hairy roots of carrots and cauliflower, induced by infection with *A. rhizogenes,* can be regenerated easily into plants, since these hairy roots have a T-region that is homologous with gene 1 coding for shoot suppression in octopine and nopaline tumors, but do not contain a gene equivalent to gene 4 (Willmitzer *et al.*, 1982b).

The question can be asked whether the general functional organization of the T-DNA-linked genes, as studied for octopine and nopaline crown gall tumors, also applies to other types of tumors or transformations induced by agrobacteria. Agropine (previously null-type) tumors (Guyon *et al.*, 1980) have a T-region that is homologous, on the basis of DNA–DNA hybridization studies (Drummond and Chilton, 1978), with genes coding for transcripts 5, 2, 1, and 4 of the "common" region of octopine and nopaline T-regions. No homology was found to transcripts 6a and 6b.

In the previous literature, there had been observations suggesting that the T-DNA could not pass through meiosis. Seeds obtained by self-fertilization of LpDH-positive plants, however, produced new plants with active T-DNA-linked genes, demonstrating that genes introduced into plant nuclei via the Ti plasmid can be sexually inherited (Otten *et al.*, 1981). A series of sexual crosses were therefore designed to study the transmission pattern of the T-DNA-specified genes. The results of these crosses demonstrate very convincingly that the T-DNA-linked genes (LpDH) are transmitted as a single Mendelian factor through both the pollen and the eggs of the originally transformed plant. These crosses also showed that the original transformed plant was a hemizygote containing T-DNA on only one of a pair of homologous chromosomes. By these crosses, tobacco plant homozygotes for the altered T-DNA were obtained (Otten *et al.*, 1981). When regenerants from different transformations are crossed, the two T-DNA loci segregate independently (De Greve *et al.*, 1982b). Subsequent experiments have demonstrated that mutant Ti plasmids reproducibly give rise to normal plants in tobacco, petunia, and potato. In all these cases, the plants were shown to contain and express the octopine or agropine synthase genes, or both, of the mutant T-DNA.

4. REGULATION OF THE ACTIVITY OF T-DNA-LINKED GENES

In the previous section, we have described the location and the function of a number of T-DNA-linked genes, and we have presented evidence showing that these genes are transcribed from T-DNA internal promoters. An important question is whether or not these genes can be regulated in their expression. A number of preliminary observations bear on this question.

The opine synthase genes appear to be essentially constitutive, since they were found to be active not only in crown gall tumors but also in the differentiated tissues of octopine-positive and nopaline-positive plants regenerated from T-DNA-containing plant cells (De Greve *et al.*, 1982b; Otten *et al.*, 1981).

On the other hand, it would appear that the root-suppressing gene 4, and possibly also the products of genes 6a and 6b, are subject to control. This conclusion was reached on the basis of grafting experiments involving teratoma tumors containing a deletion in the "common" T-region inactivating genes 5, 2, and 1 (Leemans *et al.*, 1982; Wöstemeyer *et al.*, 1983).

Cells containing such partial T-regions form green tumors from which abnormal teratoma shoots proliferate. These shoots remain small and cannot form roots, even when separated from the tumor. Such shoots were repeatedly grafted onto normal tobacco stems, which resulted in a "normalization," since the grafted shoots developed and finally produced fairly normal shoots with fertile flowers.

In contrast to normal tobacco leaves, leaf fragments of these grafted plants were able to grow when placed on hormone-free medium. Also, different parts of the grafted teratomas (stem, leaf, flower) retained the opine marker. The observations that flowers on such LpDH-positive grafts are often male-sterile (Wullems *et al.*, 1981) was wrongly considered to represent a T-DNA-controlled property. It is more likely due to a property of the SR-1 line itself, since male sterility was also observed in control plants (our observations). Moreover, in our experiments, one of the two LpDH-positive grafts investigated was found to be male-sterile, whereas the other one was completely fertile.

After selfing of this fertile graft, two types of germinating plantlets were found: normally rooting plants without LpDH activity and rootless plantlets each of which contained high levels of LpDH activity.

When these rootless plantlets were cultured *in vitro* on hormone-free medium, they formed callus at their base and developed into the type of teratomalike tissue from which the grafts originated. When the fertile LpDH-positive graft was crossed with normal SR-1 pollen, 170 seedlings were normal and LpDH-negative, 191 were rootless and LpDH-positive. After selfing, this ratio was 38:110. The results indicate that in the original graft, the LpDH gene and the gene or genes preventing root formation were 100% linked and present in a hemizygous state.

T-DNA-derived transcripts were analyzed by hybridizing polynucleotide-kinase-labeled poly(A)-containing RNA isolated from polysomes to Southern blots containing fragments of the T-DNA corresponding to genes 4, 6a, 6b, and 3. The analysis showed that the concentrations of transcripts 4 and 6 were very low in leaf tissue, whereas they were clearly present in callus tissue obtained when the leaf tissue of the grafted material was brought into tissue culture and forced to dedifferentiate into callus tissue by the addition of 2 mg/ml naphthalene acetic acid and 0.2 mg/ml kinetin or when the teratoma tissues grown on medium without hormones were used for RNA analysis. It is not known whether this correlation between the lack of tissue organization and the levels of messenger(s) 4 (and possibly 6) is the result of transcription regulation or reflects events taking place at the level of messenger stability, processing, or transport into polysomes. It is, however, important to note that the level of transcription of the LpDH gene (gene 3) was essentially the same in all these tissues.

The presence of auxin and cytokinin at the concentrations used apparently does not inhibit the expression of genes 4 and 6, nor does the absence of hormones.

These observations very strongly support our previous conclusions that the activity of gene 4 results in the suppression of root formation. In addition, the effects of this gene can be eliminated by grafting, and in this case no transcripts can be observed. Both the effect of the gene and its transcription are reestablished when leaf tissue is put on hormone-free medium and develops into teratoma tissue and, remarkably, in seedlings germinating from seeds containing gene 4 in their genome.

It is therefore obvious that the activity of gene 4 is directly involved in prevention of root formation, but can be controlled by intrinsic plant factors. Work is in progress to analyze the nature of the plant factors that control the activity of the root-suppressing gene.

5. GENERAL CONCLUSIONS

Ti plasmids of agrobacteria have provided us with an unexpected and extraordinarily potent system to study the genetic control of plant organ development. Indeed, we found that these plasmids contain DNA sequences that are transferred and integrated into chromosomes of plant cells. These so-called T-DNA sequences contain genes that are expressed in plant cells. Some of these genes make products that specifically suppress either shoot or root formation.

These observations indicate that plants have separate genetic programs for shoot development and for root development and that both these programs must be internally coordinated, since they can be controlled by the products of single genes.

Furthermore, the T-DNA remarkably has two independent but complementary mechanisms to suppress plant organ development. Each of these mechanisms alone may be sufficient to suppress the development of transformed cells. One of these mechanisms appears to function via a growth-hormone-like mechanism and therefore suppresses both transformed and untransformed cells, thus explaining why uncloned primary tumors, consisting of a mixture of T-DNA-containing and normal untransformed cells, nevertheless grow as a uniformly undifferentiated callus. This mechanism might well play a role in the production of auxin- and cytokininlike hormones. These observations are therefore consistent with the observations of Amasino and Miller (1982), who studied the endogenous levels of auxin and cytokinin in teratoma and unorganized tobacco and found them to be similar to those needed to artificially induce these growth patterns in normal cells. The other mechanism is even more intriguing, since it appears to suppress transformed plant cells only.

REFERENCES

Amasino, R. M., and Miller, C. O., 1982, *Plant Physiol.* **69:**389–392.

Barton, K. A., Binns, A. N., Matzke, A. J. M., and Chilton, M.-D., 1983, *Cell* **32:**1033–1043.

Bomhoff, G., Klapwijk, P. M., Kester, H. C. M., Schilperoort, R. A., Hernalsteens, J. P., and Schell, J., 1976, *Mol. Gen. Genet.* **145:**177–181.

Chilton, M.-D., Drummond, M. H., Merlo, D. J., Sciaky, D., Montoya, A. L., Gordon, M. P., and Nester, E. W., 1977, *Cell* **11:**263–271.

Chilton, M.-D., Drummond, M. H., Merlo, D. J., and Sciaky, D., 1978, *Nature* **275:**147–149.

Chilton, M.-D., Saiki, R. K., Yadav, N., Gordon, M. P., and Quetier, F., 1980, *Proc. Natl. Acad. Sci. U.S.A.* **77:**4060–4064.

Currier, T. C., and Nester, E. W., 1976, *J. Bacteriol.* **126:**157–165.

De Beuckeleer, M., De Block, M., De Greve, H., Depicker, A., De Vos, R., De Vos, G., De Wilde, M., Dhaese, P., Dobbelaere, M. R., Engler, G., Genetello, C., Hernalsteens, J. P., Holsters, M., Jacobs, A., Schell, J., Seurinck, J., Silva, B., Van Haute, E., Van Montagu, M., Van Vliet, F., Villarroel, R., and Zaenen, I., 1978, in: *Proceedings of the IVth International Conference on Plant Pathogenic Bacteria* (M. Ridé, ed.), INRA, Angers, France, pp. 115–126.

De Beuckeleer, M., Lemmers, M., De Vos, G., Willmitzer, L., Van Montagu, M., and Schell, J., 1981, *Mol. Gen. Genet.* **183:**283–288.

De Greve, H., Decraemer, H., Seurinck, J., Van Montagu, M., and Schell, J., 1981, *Plasmid* **6:**235–248.

De Greve, H., Dhaese, P., Seurinck, J., Lemmers, M., Van Montagu, M., and Schell, J., 1982a, *J. Mol. Appl. Genet.* **1**:499–512.

De Greve, H., Leemans, J., Hernalsteens, J. P., Thia-Toong, L., De Beuckeleer, M., Willmitzer, L., Otten, L., Van Montagu, M., and Schell, J., 1982b, *Nature (London)* **300**:752–755.

Depicker, A., Van Montagu, M., and Schell, J., 1978, *Nature (London)* **275**:150–153.

Depicker, A., Stachel, S., Dhaese, P., Zambryski, P., and Goodman, H. M., 1982, *J. Mol. Appl. Genet.* **1**:561–574.

Drummond, M. H., and Chilton, M.-D., 1978, *J. Bacteriol.* **136**:1178–1183.

Drummond, M. H., Gordon, M. P., Nester, E. W., and Chilton, M.-D., 1977, *Nature (London)* **269**:535–536.

Ellis, J. G., Kerr, A., Tempé, J., and Petit, A., 1979, *Mol. Gen. Genet.* **173**:263–269.

Engler, G., Holsters, M., Van Montagu, M., Schell, J., Hernalsteens, J. P., and Schilperoort, R. A., 1975, *Mol. Gen. Genet.* **138**:345–349.

Engler, G., Depicker, A., Maenhaut, R., Villarroel-Mandiola, R., Van Montagu, M., and Schell, J., 1981, *J. Mol. Biol.* **152**:183–208.

Firmin, J. L., and Fenwick, G. R., 1978, *Nature (London)* **276**:842–844.

Fitzgerald, M., and Shenk, T., 1981, *Cell* **24**:251–260.

Garfinkel, D. J., Simpson, R. B., Ream, L. W., White, F. F., Gordon, M. P., and Nester, E. W., 1981, *Cell* **27**:143–153.

Gelvin, S. B., Gordon, M. P., Nester, E. W., and Aronson, A. I., 1981, *Plasmid* **6**:17–29.

Genetello, C., Van Larebeke, N., Holsters, M., Depicker, A., Van Montagu, M., and Schell, J., 1977, *Nature (London)* **265**:561–563.

Gurley, W. B., Kemp, J. D., Albert, M. J., Sutton, D. W., and Callis, J., 1979, *Proc. Natl. Acad. Sci. U.S.A.* **76**:2828–2832.

Guyon, P., Chilton, M.-D., Petit, A., and Tempé, J., 1980, *Proc. Natl. Acad. Sci. U.S.A.* **77**:2693–2697.

Hernalsteens, J. P., Van Vliet, F., De Beuckeleer, M., Depicker, A., Engler, G., Lemmers, M., Holsters, M., Van Montagu, M., and Schell, J., 1980, *Nature (London)* **287**: 654–656.

Holsters, M., De Waele, D., Depicker, A., Messens, E., Van Montagu, M., and Schell, J., 1978, *Mol. Gen. Genet.* **163**:181–187.

Holsters, M., Villarroel, R., Gielen, J., Seurinck, J., De Greve, H., Van Montagu, M., and Schell, J., 1983, *Mol. Gen. Genet.* **190**:35–41.

Hooykaas, P. J. J., Klapwijk, P. M., Nuti, M. P., Schilperoort, R. A., and Rörsch, A., 1977, *J. Gen. Microbiol.* **98**:477–484.

Joos, H., Inzé, D., Caplan, A., Sormann, M., Van Montagu, M., and Schell, J., 1983, *Cell* **32**:1057–1067.

Kerr, A., 1969, *Nature (London)* **223**:1175–1176.

Kerr, A., 1971, *Physiol. Plant Pathol.* **1**:241–246.

Kerr, A., Manigault, P., and Tempé, J., 1977, *Nature (London)* **265**:560–561.

Klapwijk, P. M., Scheulderman, T., and Schilperoort, R. A., 1978, *J. Bacteriol.* **136**:775–785.

Koekman, B. P., Ooms, G., Klapwijk, P. M., and Schilperoort, R. A., 1979, *Plasmid* **2**:347–357.

Leemans, J., Shaw, C., Deblaere, R., De Greve, H., Hernalsteens, J. P., Maes, M., Van Montagu, M., and Schell, J., 1981, *J. Mol. Appl. Genet.* **1**:149–164.

Leemans, J., Deblaere, R., Willmitzer, L., De Greve, H., Hernalsteens, J. P., Van Montagu, M., and Schell, J., 1982, *Eur. Mol. Biol. Assoc. J.* **1**:147–152.

Lemmers, M., De Beuckeleer, M., Holsters, M., Zambryski, P., Depicker, A., Hernalsteens, J. P., Van Montagu, M., and Schell, J., 1980, *J. Mol. Biol.* **144**:353–376.

Merlo, D. J., and Nester, E. W., 1977, *J. Bacteriol.* **129**:76–80.

Merlo, D. J., Nutter, R. C., Montoya, A. L., Garfinkel, D. J., Drummond, M. H., Chilton, M.-D., Gordon, M. P., and Nester, E. W., 1980, *Mol. Gen. Genet.* **177**:637–643.

Morris, R. O., Akiyoshi, D. E., MacDonald, E. M. S., Morris, J. W., Regier, D. A., and Zaerr, J. B., 1982, in: *Plant Growth Substances 1982* (P. F. Wareing, ed.), Academic Press, London, pp. 175–183.

Ooms, G., Hooykaas, P. J., Moleman, G., and Schilperoort, R. A., 1981, *Gene* **14**:33–50.

Otten, L., De Greve, H., Hernalsteens, J. P., Van Montagu, M., Schieder, O., Straub, J., and Schell, J., 1981, *Mol. Gen. Genet.* **183**:209–213.

Petit, A., Tempé, J., Kerr, A., Holsters, M., Van Montagu, M., and Schell, J., 1978a, *Nature (London)* **271**:570–572.

Petit, A., Dessaux, Y., and Tempé, J., 1978b, in: *Proceedings of the IVth International Conference on Plant Pathogenic Bacteria* (M. Ridé, ed.), INRA, Angers, France, pp. 143–152.

Schell, J., 1975, in: *Genetic Manipulations with Plant Materials* (L. Ledoux, ed.), Plenum Press, New York, pp. 163–181.

Schell, J., and Van Montagu, M., 1977, *Brookhaven Symp. Biol.* **29**:36–49.

Schell, J., Van Montagu, M., De Beuckeleer, M., De Block, M., Depicker, A., De Wilde, M., Engler, G., Genetello, C., Hernalsteens, J. P., Holsters, M., Seurinck, J., Silva, B., Van Vliet, F., and Villarroel, R., 1979, *Proc. R. Soc. Lond. Ser. B* **204**:251–266.

Schröder, G., and Schröder, J., 1982, *Mol. Gen. Genet.* **185**:51–55.

Schröder, J., Schröder, G., Huisman, H., Schilperoort, R. A., and Schell, J., 1981a, *FEBS Lett.* **129**:166–168.

Schröder, J., Hillebrandt, A., Klipp, W., and Pühler, A., 1981b, *Nucleic Acids Res.* **9**:5187–5202.

Simpson, R. B., O'Hara, P. J., Krook, W., Montoya, A. L., Lichtenstein, C., Gordon, M. P., and Nester, E. W., 1982, *Cell* **29**:1005–1014.

Skoog, F., and Miller, C. O., 1957, *Symp. Soc. Exp. Biol.* **11**:118–131.

Thomashow, M. F., Nutter, R., Montoya, A. L., Gordon, M. P., and Nester, E. W., 1980a, *Cell* **19**:729–739.

Thomashow, M. F., Nutter, R., Postle, K., Chilton, M.-D., Blattner, F. R., Powell, A., Gordon, M. P., and Nester, E. W., 1980b, *Proc. Natl. Acad. Sci. U.S.A.* **77**:6448–6452.

Van Larebeke, N., Engler, G., Holsters, M., Van den Elsacker, S., Zaenen, I., Schilperoort, R. A., and Schell, J., 1974, *Nature (London)* **252**:169–170.

Van Larebeke, N., Genetello, C., Schell, J., Schilperoort, R. A., Hermans, A. K., Hernalsteens, J. P., and Van Montagu, M., 1975, *Nature (London)* **255**:742–743.

Van Larebeke, N., Genetello, C., Hernalsteens, J. P., Depicker, A., Zaenen, I., Messens, E., Van Montagu, M., and Schell, J., 1977, *Mol. Gen. Genet.* **152**:119–124.

Van Montagu, M., and Schell, J., 1979, in: *Plasmids of Medical, Environmental and Commercial Importance* (K. Timmis and A. Pühler, eds.), Elsevier, Amsterdam, pp. 71–96.

Watson, B., Currier, T. C., Gordon, M. P., Chilton, M.-D., and Nester, E. W., 1975, *J. Bacteriol.* **123**:255–264.

Willmitzer, L., De Beuckeleer, M., Lemmers, M., Van Montagu, M., and Schell, J., 1980, *Nature (London)* **287**:359–361.

Willmitzer, L., Otten, L., Simons, G., Schmalenbach, W., Schröder, J., Schröder, G., Van Montagu, M., De Vos, G., and Schell, J., 1981a, *Mol. Gen. Genet.* **182**:255–262.

Willmitzer, L., Schmalenbach, W., and Schell, J., 1981b, *Nucleic Acids Res.* **9**:4801–4812.

Willmitzer, L., Simons, G., and Schell, J., 1982a, *Eur. Mol. Biol. Assoc. J.* **1**:139–146.

Willmitzer, L., Sanchez-Serrano, J., Buschfeld, E., and Schell, J., 1982b, *Mol. Gen. Genet.* **186**:16–22.

Willmitzer, L., Dhaese, P., Schreier, P. H., Schmalenbach, W., Van Montagu, M., and Schell, J., 1983, *Cell* **32**:1045–1056.

Wöstemeyer, A., Otten, L., De Greve, H., Hernalsteens, J. P., Leemans, J., Van Montagu, M., and Schell, J., 1983, in: *Genetic Engineering in Eukaryotes* (P. Lurquin and A. Kleinhofs, eds.), Plenum Press, New York, pp. 137–151.

Wullems, G. J., Molendijk, L., Ooms, G., and Schilperoort, R. A., 1981, *Cell* **24:**719–727.

Yadav, N. S., Postle, K., Saiki, R. K., Thomashow, M. F., and Chilton, M.-D., 1980, *Nature (London)* **287:**458–461.

Zaenen, I., Van Larebeke, N., Teuchy, H., Van Montagu, M., and Schell, J., 1974, *J. Mol. Biol.* **86:**109–127.

Zambryski, P., Holsters, M., Kruger, K., Depicker, A., Schell, J., Van Montagu, M., and Goodman, H. M., 1980, *Science* **209:**1385–1391.

Zambryski, P., Depicker, A., Kruger, K., and Goodman, H., 1982, *J. Mol. Appl. Genet.* **1:**361–370.

10

Activity of a Chick Collagen Gene in Heterologous and Homologous Cell-Free Extracts

GLENN T. MERLINO, JAYA SIVASWAMI TYAGI, BENOIT DE CROMBRUGGHE, and IRA PASTAN

1. INTRODUCTION

The collagens are a family of proteins found in most animal tissues, where they play various important structural roles as extracellular matrix constituents. Type I collagen, which consists of one $\alpha2$ and two $\alpha1$ polypeptide chains, is a prevalent protein in bone, tendon, and skin (Ramachandran and Reddi, 1976) and is enriched in cultured chick embryo fibroblasts (CEF). We have used CEF in culture as a model system to study the regulation of type I collagen synthesis. The gene coding for $\alpha2$ (type I) collagen, isolated in a series of overlapping genomic clones, is approximately 38 kilobases long (Ohkubo *et al.*, 1980; Vogeli *et al.*, 1981; Wozney *et al.*, 1981). Figure 1 shows that the coding information in this gene is subdivided into more than 50 exons. The transcription start site at the extreme 5' end of the $\alpha2$ collagen gene has been located (Vogeli *et al.*, 1981) and subcloned into the plasmid pBR322 (see Section 2). Residing upstream of the start site ($+1$) are a TATA

GLENN T. MERLINO, JAYA SIVASWAMI TYAGI, BENOIT DE CROMBRUGGHE, and IRA PASTAN • Laboratory of Molecular Biology, National Cancer Institute, National Institutes of Health, Bethesda, Maryland 20205.

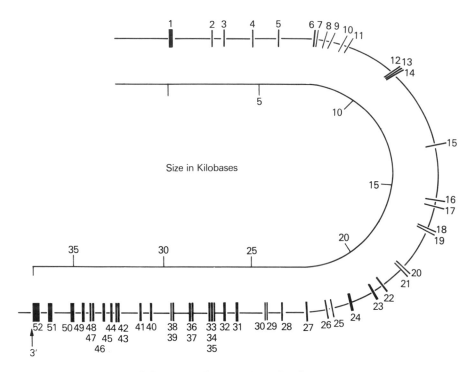

FIGURE 1.   Chick α2 (type I) collagen gene.

box (-33), a CAT box (-84), and three inverted repeat sequences that have
the potential to form hairpin-loop structures (Fig. 2). The sequences involved
in the formation of these three dyads of symmetry overlap and are thus
mutually exclusive. Depending on which structure is formed, the CAT box
is localized in different regions of the hairpin (Fig. 2, A–C). These sequences
may prove to have regulatory significance.

Transcriptional control represents a critical element in the regulation of
gene expression. A thorough understanding of the nature of this control must
ultimately involve reproduction of cellular transcription events in reconstituted
cell-free systems. Recently, significant progress has been made toward developing such *in vitro* systems (Weil *et al.*, 1979; Manley *et al.*, 1980). One
of these, described by Manley, utilizes an ammonium-sulfate-concentrated
extract derived from whole HeLa cells in concert with a viral DNA template.
Many class II eukaryotic genes have now been studied using this HeLa transcription system (i.e., Wasylyk *et al.*, 1980; Talkington *et al.*, 1980; Tsai *et*

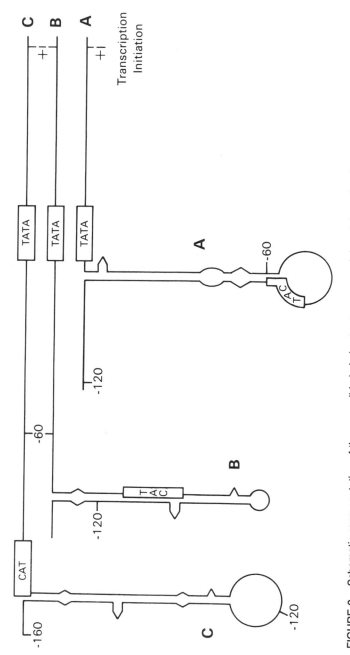

FIGURE 2. Schematic representation of three possible hairpin structures around the α2 (type I) collagen gene promoter region.

al., 1981). More recently, homologous cell-free systems have been described for the insects *Bombyx* and *Drosophila*. Tsuda and Suzuki (1981) found that transcription of the fibroin gene was correctly initiated when used in crude extracts derived from silk glands. Craine and Kornberg (1981) have used cytoplasmic extracts from heat-shocked *Drosophila* cells to induce inactive heat-shock genes in isolated nuclei.

We first wanted to determine whether the $\alpha2$ (I) collagen gene could be accurately transcribed *in vitro* using the HeLa system described by Manley *et al.* (1980). This accomplished, we wished to examine transcription of the chick $\alpha2$ (I) collagen gene in extracts derived from both normal and Rous sarcoma virus (RSV)-transformed CEF because transformation by this virus

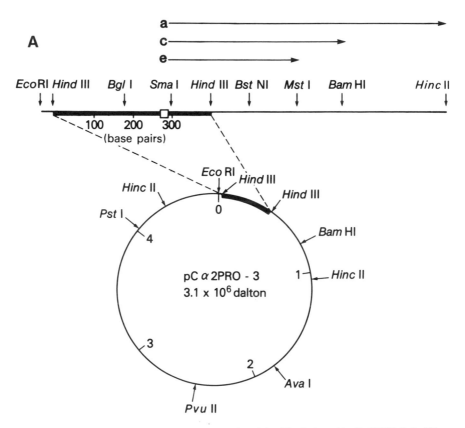

FIGURE 3. Structure (A) and transcriptional activity (B) of plasmid pCα2PRO-3. In (A), the heavy line represents the 400-bp chick DNA fragment and the white box represents the promoter region. In (B), ^{32}P-labeled transcripts in lanes a, c, and e were generated

results in a striking reduction in the synthesis of both the $\alpha 2$ (I) collagen protein and the RNA encoding $\alpha 2$ (I) collagen (Green *et al.*, 1966; Levinson *et al.*, 1975; Hata and Peterkofsky, 1977; Adams *et al.*, 1977, 1979; Howard *et al.*, 1978; Rowe *et al.*, 1978; Sobel *et al.*, 1981; Avvedimento *et al.*, 1981). The latter suggests that RSV transformation mediates the expression of this collagen gene in CEF by a transcriptional control mechanism. Here, we describe accurate transcription of the chick $\alpha 2$ (I) collagen gene by homologous RSV-transformed CEF extracts. In these extracts, specific transcription can be detected only by using *in vitro*-specific S1 nuclease mapping and avian myeloblastosis virus (AMV) reverse-transcriptase-catalyzed primer extension. These techniques have been used to demonstrate accurate tran-

B   FRACTIONATION OF *IN VITRO*
TRANSCRIPTS ON 4% ACRYLAMIDE

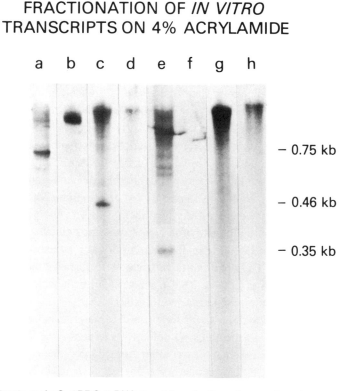

by using HeLa extracts and pCα2PRO-3 DNA template cut with various endonucleases [a, c, and e arrows in (A)]; (b, d, f) plus α-amanitin; (g) pCα2PRO-3 × *Eco*RI; (h) linearized pBR322.

scription of the collagen gene in extracts derived from Chinese hamster ovary (CHO) cells as well.

2. COLLAGEN-GENE TRANSCRIPTION IN HETEROLOGOUS HeLa WHOLE-CELL EXTRACTS

The *in vivo* transcription start of the α2 (I) collagen gene was located by a variety of methods within a 400-base-pair (bp) *Hinf*I genomic DNA fragment (Vogeli *et al.*, 1981). This fragment was subcloned into the *Hind*III site of plasmid pBR322 using *Hind*III decanucleotide linkers (Merlino *et al.*, 1981). This plasmid (pCα2PRO-3) was then used as template for transcription by HeLa-cell extracts prepared as described elsewhere (Manley *et al.*, 1980; Merlino *et al.*, 1981). The runoff transcription strategy used for determining the start of *in vitro* transcription is shown in Fig. 3A. Plasmid pCα2PRO-3 was truncated at various sites downstream from the putative start site using three different restriction enzymes (Fig. 3A, arrows a, c, and e). Transcription of these truncated DNA templates resulted in ^{32}P-labeled RNA transcripts of three different sizes (Fig. 3B, lanes a, c, and e), all of which were sensitive to 1 μg/ml α-amanitin (lanes b, d, and f), indicating that they were products of RNA polymerase II. The sizes of these three transcripts were consistent with their having the same point of origin.

That this was the correct transcription start site was confirmed by comparing the sequence of the *in vitro*-made RNA transcripts with the structure of the DNA templates and by identifying the first and second bases of *in vitro*-synthesized RNA transcripts by determining which nucleotides were associated with the cap (Merlino *et al.*, 1981). This *in vitro* start site is the same as the initiation site of *in vivo*-synthesized chick collagen RNA (Vogeli *et al.*, 1981). In addition, the transcriptional efficiency of the collagen promoter in the HeLa extract is comparable to that of other known promoters: at least as strong as the RSV promoter, but somewhat weaker than the adenovirus promoter (Merlino *et al.*, 1981).

The HeLa-cell extracts were useful for demonstration of strong, accurate transcription off the chick α2 collagen gene template. However, as with any system using heterologous combinations of polymerase, factors, and DNA template, the HeLa extracts were limited in their ability to reproduce transcriptional control events. We therefore decided to use RSV-transformed CEF to prepare extracts because large amounts of these cells were readily available and because others have had success using extracts derived from transformed cell types (Weil *et al.*, 1979; Manley *et al.*, 1980). The resulting system could be used to examine the activity of the chick α2 collagen gene in an extract containing chicken-specific RNA polymerase and regulatory factors.

3. COLLAGEN-GENE TRANSCRIPTION IN HOMOLOGOUS CHICKEN WHOLE-CELL EXTRACTS

3.1. Detection by Modified S1 Nuclease Mapping

During preliminary experiments, we attempted to analyze *in vitro*-synthesized RNA from both normal and RSV-transformed CEF extracts on denaturing gels by runoff transcription techniques that were successful with HeLa extracts using a truncated DNA template. However, only a smear of ^{32}P-labeled RNA products was observed (Fig. 4, lane b) (Merlino *et al.*, 1982). The majority of this activity was resistant to both 1 and 200 μg/ml α-amanitin,

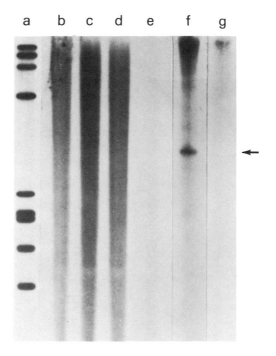

FIGURE 4. Runoff transcription of truncated collagen gene plasmid DNA (× *Bam*HI) using RSV-CEF whole-cell extracts. Lanes: (a) φ× markers; (b) RSV-CEF extracts; (c) RSV-CEF + 1 μg/ml α-amanitin; (d) RSV-CEF + 200 μg/ml α-amanitin; (e) RSV-CEF + 50 μg/ml actinomycin D; (f) HeLa extracts; (g) HeLa + 1 μg/ml α-amanitin.

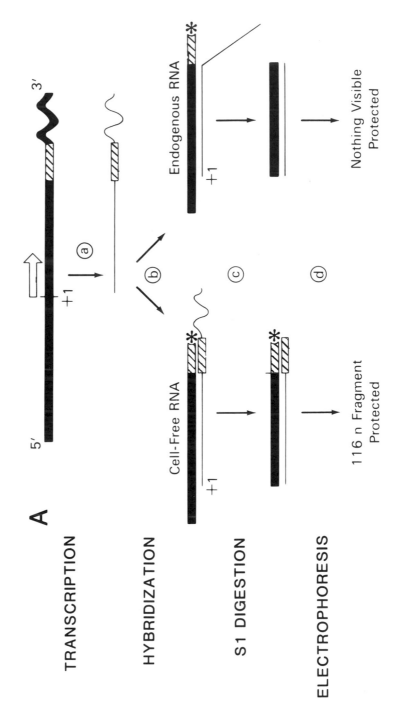

HYBRIDIZATION

S1 DIGESTION

ELECTROPHORESIS

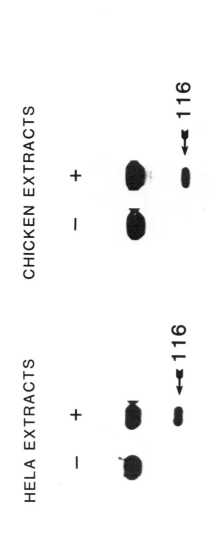

FIGURE 5. (A) Schematic representation of modified, *in vitro*-specific S1 nuclease mapping. Heavy lines are chick DNA; wavy lines, pBR322; thin lines, RNA; hatched boxes, *Hind*III linkers. (*) ^{32}P label. (B) Autoradiograph of resulting DNA fragments with (+) and without (−) added template DNA.

suggesting that it was the result of transcription by RNA polymerase I (lanes c and d). This was in marked contrast to HeLa-cell extracts, which exhibited little background activity (lanes f and g).

To detect specific RNA-polymerase-II-catalyzed transcription within these chicken extracts, we decided to employ the more sensitive S1 nuclease mapping procedure (Berk and Sharp, 1977). For clearer results, it is desirable to use conditions under which the DNA probe is labeled at one end and hybridized to unlabeled RNA (Weaver and Weissman, 1979). The drawback of this approach is that the DNA probe cannot distinguish between the newly made *in vitro*-synthesized RNA transcripts and endogenous *in vivo* RNA. To circumvent this problem, we have modified the S1 nuclease mapping procedure by utilizing a DNA probe that is end-labeled at the *Hind*III linker used to subclone the collagen DNA into pBR322 (Merlino *et al.*, 1981). This procedure is schematized in Fig. 5A.

The collagen-promoter-containing plasmid pCα2PRO-3 was used as template. Extracts were made from RSV-transformed CEF (Merlino *et al.*, 1982) using a modification of the procedure of Manley *et al.* (1980). RNAs, both endogenous and cell-free, were isolated and hybridized to a 158-base *Ava*II–*Hind*III fragment (5′ labeled at the *Hind*III site), treated with S1 nuclease, and subjected to denaturing 7% polyacrylamide gel electrophoresis. By hybridizing to this site, only *in vitro*-synthesized RNA should be capable of protecting the [^{32}P]-DNA probe from S1 nuclease degradation. *In vivo* collagen RNA, therefore, should not be visualized. Figure 5B shows that a single-stranded DNA fragment of 116 bases was protected when pCα2PRO-3 was used as template. This DNA fragment comigrated with the DNA fragment generated by the same DNA template in HeLa extracts (Fig. 5).

We have determined that the start site of transcription of the collagen gene is identical in both chicken and HeLa extracts by electrophoresing the S1-nuclease-protected DNA fragment on a denaturing polyacrylamide gel adjacent to a Maxam and Gilbert (1977) sequence latter of the 158-base mapping probe (Merlino *et al.*, 1982). In addition, we have used this S1 approach to determine that the adenovirus 2 major late promoter (in plasmid pSmaF, from T. Weil) also functions accurately in the RSV-transformed CEF extracts (Merlino *et al.*, 1982).

3.2. Detection by Primer Extension

Although the S1 nuclease mapping procedure represents a powerful approach to the study of low-level transcription, it is subject to possible artifactual degradation events at AT-rich regions (Hansen *et al.*, 1981). We therefore determined whether a modification of AMV reverse-transcriptase-catalyzed primer extension (Wickens *et al.*, 1978) could be used to confirm accurate homologous *in vitro* transcription. Figure 6 illustrates the strategy used for *in vitro*-specific primer extension. A 132-base *Rsa*I–*Hind*III pBRf322-specific DNA primer labeled at the 5′ end was hybridized to RNA synthesized in chicken extracts off the α2 collagen gene template. The pBR322 primer could not hybridize to endogenous RNA, only to RNA made off the pBR322-containing pCα2PRO-3 plasmid template. AMV reverse transcriptase was used to extend the DNA primer to the end of the cell-free RNA, and the resulting DNAs were electrophoresed on a denaturing 5.5% polyacrylamide gel (for details, see Merlino *et al.*, 1982).

Figure 7 confirms that both HeLa and chicken extracts accurately initiated the synthesis of collagen RNA (lanes a and c). Transcription is completely

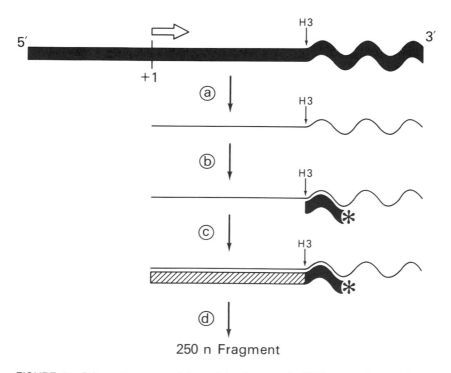

FIGURE 6. Schematic representation of *in vitro*-specific AMV reverse-transcriptase-catalyzed primer extension. Heavy lines are chick DNA; wavy lines, pBR322; thin lines, RNA; hatched lines, reverse-transcribed DNA. (*) ³²P label; (H3) *Hind*III linker.

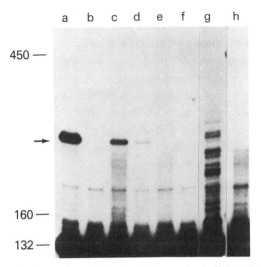

FIGURE 7.   Primer extension using cell-free RNA transcribed off the α2 collagen-gene template. Lanes: (a) HeLa extract; (b) HeLa + 200 μg/ml α-amanitin; (c) RSV-CEF extract + 5 mM creatine phosphate; (d) RSV-CEF; (e) RSV-CEF + 200 μg/ml α-amanitin; (f) RSV-CEF + 50 μg/ml actinomycin D; (g) naked DNA + partially purified calf thymus RNA polymerase II; (h) overexposure of (e). From Merlino *et al.* (1982).

inhibited by α-amanitin in both extracts (lanes b and e). To determine whether specific factors were necessary for the accurate transcription displayed in Fig. 7, the naked α2 collagen gene template pCα2PRO-3 was transcribed using partially purified calf thymus RNA polymerase II (Tyagi *et al.*, 1982). Figure 7 (lane g) reveals that initiation of transcription occurred at a large number of sites, but not significantly from the correct *in vivo–in vitro* transcription start site. This result clearly shows that chicken factors or HeLa factors are necessary for accurate α2 collagen-gene transcription.

4.   OPTIMIZATION OF THE CHICKEN TRANSCRIPTION SYSTEM

We next wanted to optimize the RSV-transformed CEF *in vitro* transcription system. It was found that 2–10 mM creatine phosphate stimulated specific collagen transcription 5- to 10-fold in agreement with the results of Handa *et al.* (1981). Surprisingly, high concentrations of GTP and UTP (2–5 mM) were also found to enhance RNA-polymerase-II-catalyzed collagen-gene transcription by a factor of 5–10 (Fig. 8) (Merlino *et al.*, 1982). Higher concentrations of these ribonucleoside triphosphates did not stimulate, nor did ATP or CTP. These results were obtained by quantitating autoradiographic data from both S1 nuclease mapping and primer extension experiments by microdensitometry.

Because the molar concentrations of GTP and UTP were much higher than those expected to be sufficient for RNA polymerization, it is conceivable that these nucleotides were acting *in vitro* at some other level. Phosphorylation is certainly one candidate for a mechanism by which nucleotides can alter

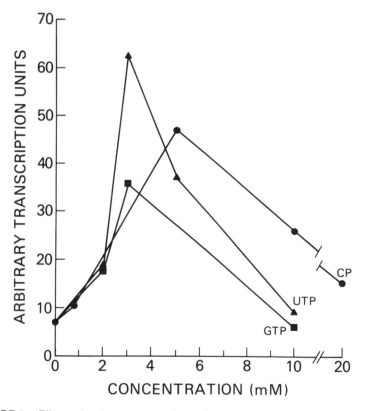

FIGURE 8.   Effects of various concentrations of creatine phosphate, GTP, and UTP on collagen-gene transcription by RSV-CEF extracts. See the text for the method of quantitation.

macromolecular function. There is considerable support for the notion that a phosphorylation–dephosphorylation mechanism plays a role in regulating RNA polymerase activity both *in vivo* and *in vitro* (i.e., Dahmus, 1981). Whether the observed stimulation is truly related to such a mechanism remains to be determined.

5. APPLICATION OF MODIFIED S1 NUCLEASE MAPPING AND PRIMER EXTENSION TECHNIQUES

Because we had been successful in detecting accurate transcription in RSV-transformed CEF extracts, we examined whether modified S1 nuclease mapping and primer extension would be of value in analyzing other *in vitro* transcription systems. The activity of the chick α2 collagen gene was therefore examined in a heterologous extract derived from CHO cells. When RNA synthesized in cell-free CHO extracts was analyzed by AMV reverse-tran-scriptase-catalyzed primer extension as described in Section 3.2, it was found that these extracts were quite capable of accurately transcribing the chicken collagen gene (Merlino *et al.*, 1982). However, analysis of CHO extracts by runoff transcription revealed only a smear of activity similar to that observed for both normal and transformed CEF extracts. These two techniques may therefore be useful for analyzing specific transcription in other types of cell-free extracts.

We also attempted to detect accurate transcriptional initiation of the α2 collagen gene in extracts derived from normal CEF using these techniques. When cell-free RNA synthesized by normal CEF extracts was analyzed by primer extension, no 250-base extended DNA fragment was formed, indicating that accurate transcription was not occurring (Merlino *et al.*, 1982). However, the CEF extracts were not deficient in RNA polymerase II activity. There must be other reasons for the failure of normal CEF to exhibit accurate *in vitro* transcription.

Groudine and Weintraub (1980) have found that RSV transformation of CEF results in RNA accumulation from about 1000 new transcription units. It is possible that normal CEF keep much tighter control over *in vivo* RNA polymerase II transcription that do RSV-transformed CEF. In addition, this gross difference in gene regulation may be mirrored *in vitro*. If this is true, it is not surprising that most of the extracts demonstrated to accurately initiate eukaryotic gene transcription were derived from the transformed cell types: HeLa, KB, CHO, and RSV-transformed CEF (Manley *et al.*, 1980; Weil *et al.*, 1979; Merlino *et al.*, 1982).

We are at present using the *in vitro* transcription systems described above to identify and characterize factors involved in the regulation of the chick α2 (I) collagen gene. We have identified in chicken embryos and cultured chicken cells a factor that blocks the accumulation of RNA polymerase II transcripts *in vitro* (Tyagi *et al.*, 1982). The relationship between this inhibitor and gene regulation in chicken cells is at present under investigation.

6. SUMMARY

Cloned chick genomic DNA containing the 5′ end of the α2 (type I) collagen gene was initially used as template in the heterologous *in vitro* HeLa-cell transcription system described by Manley *et al.* (1980). By sizing *in vitro*-synthesized RNA runoff transcription products, it was found that RNA polymerase II initiates from a specific site that is the same site as that from which *in vivo*-synthesized collagen RNA starts. Transcription was found to initiate 33 bp downstream from a Goldberg–Hogness (TATA) sequence and 84 bp downstream from a CAT box. When runoff transcription methods failed to detect accurate transcription of the α2 (type I) collagen gene in a homologous RSV-transformed CEF extract, we made use of two other techniques. The first method is a modification of the S1 nuclease mapping procedure, which utilizes a DNA probe labeled with ^{32}P at the 5′ end of the *Hind*III linker originally used to clone the collagen promoter region into pBR322. The probe distinguishes newly made, specific RNA from endogenous RNA and nonspecific transcripts. Using this procedure, we have found that chicken whole-cell extracts support accurate initiation of transcription of the chicken α2 (type I) collagen DNA template. Addition of either creatine phosphate, GTP, or UTP to concentrations of about 3–5 mM was found to stimulate RNA polymerase II transcription by 5- to 10-fold. The second method employs an AMV reverse-transcriptase-catalyzed primer extension procedure, rendered *in vitro*-specific by use of a pBR322 fragment as primer. These techniques have been used to demonstrate accurate transcription of this collagen gene in extracts derived from CHO cells as well and may be useful for analyzing specific transcription in other types of cell-free extracts.

ACKNOWLEDGMENTS. The authors wish to thank R. Coggin for typing this manuscript and R. Steinberg for preparing the photographs. G.T.M. was supported by a postdoctoral fellowship from the Arthritis Foundation.

REFERENCES

Adams, S. L., Sobel, M. E., Howard, B. H., Olden, K., Yamada, K. M., de Crombrugghe, B., and Pastan, I., 1977, *Proc. Natl. Acad. Sci. U.S.A.* **74:**3399–3403.

Adams, S. L., Alwine, J. S., de Crombrugghe, B., and Pastan, I., 1979, *J. Biol. Chem.* **254:**4935–4938.

Avvedimento, E., Yamada, Y., Lovelace, E., Vogeli, G., de Crombrugghe, B., and Pastan, I., 1981, *Nucleic Acids Res.* **9:**1123–1131.

Berk, A. J., and Sharp, P. A., 1977, *Cell* **12:**721–732.

Craine, B. L., and Kornberg, T., 1981, *Cell* **25:**671–681.

Dahmus, M. E., 1981, *J. Biol. Chem.* **256:**3332–3339.

Green, H., Todaro, G. J., and Goldberg, B., 1966, *Nature (London)* **209:**916–917.

Groudine, M., and Weintraub, H., 1980, *Proc. Natl. Acad. Sci. U.S.A.* **77:**5351–5354.

Handa, H., Kaufman, R. J., Manley, J., Gefter, M., and Sharp, P. A., 1981, *J. Biol. Chem.* **256:**478–482.

Hansen, U., Tenen, D. G., Livington, D. M., and Sharp, P. A., 1981, *Cell* **27:**603–612.

Hata, R., and Peterkofsky, B., 1977, *Proc. Natl. Acad. Sci. U.S.A.* **74:**2933–2937.

Howard, B. H., Adams, S. L., Sobel, M. E., Pastan, I., and de Crombrugghe, B., 1978, *J. Biol. Chem.* **253:**5869–5874.

Levinson, W., Bhatnager, R. S., and Lin, T.-Z., 1975, *J. Natl. Cancer Inst.* **55:**807–810.

Manley, J. L., Fire, A., Cano, A., Sharp, P. A., and Gefter, M. L., 1980, *Proc. Natl. Acad. Sci. U.S.A.* **77:**3855–3859.

Maxam, A. M., and Gilbert, W., 1977, *Proc. Natl. Acad. Sci. U.S.A.* **74:**560–564.

Merlino, G. T., Vogeli, G., Yamamoto, T., de Crombrugghe, B., and Pastan, I., 1981, *J. Biol. Chem.* **256:**11251–11258.

Merlino, G. T., Tyagi, J. S., de Crombrugghe, B., and Pastan, I., 1982, *J. Biol. Chem.* **257:**7254–7261.

Ohkubo, H., Vogeli, G., Mudryj, M., Avvedimento, V. E., Sullivan, M., Pastan, I., and de Crombrugghe, B., 1980, *Proc. Natl. Acad. Sci. U.S.A.* **77:**7059–7063.

Ramachandran, G. N., and Reddi, A. H. (eds.), 1976, *Biochemistry of Collagen,* Plenum Press, New York.

Rowe, D. W., Moen, R. C., Davidson, J. M., Byers, P. H., Bornstein, P., and Palmiter, R. D., 1978, *Biochemistry* **17:**1581–1590.

Sobel, M. E., Yamamoto, T., de Crombrugghe, B., and Pastan, I., 1981, *Biochemistry* **20:**2678–2684.

Talkington, C. A., Nishioka, Y., and Leder, P., 1980, *Proc. Natl. Acad. Sci. U.S.A.* **77:**7132–7136.

Tsai, S. Y., Tsai, M.-J., and O'Malley, B., 1981, *Proc. Natl. Acad. Sci. U.S.A.* **78:**879–883.

Tsuda, M., and Suzuki, Y., 1981, *Cell* **27:**175–182.

Tyagi, J. S., Merlino, G. T., de Crombrugghe, B., and Pastan, I., 1982, *J. Biol. Chem.* **257:**13001–13008.

Vogeli, G., Ohkubo, H., Sobel, M. E., Yamada, Y., Pastan, I., and de Crombrugghe, B., 1981, *Proc. Natl. Acad. Sci. U.S.A.* **78:**5334–5338.

Wasylyk, B., Kedinger, C., Corden, J., Brison, O., and Chambon, P., 1980, *Nature (London)* **285:**367–373.

Weaver, R. F., and Weissman, C., 1979, *Nucleic Acids Res.* **7:**1175–1193.

Weil, P. A., Luse, D. S., Segall, J., and Roeder, R. G., 1979, *Cell* **18:**469–484.

Wickens, M. P., Buell, G. N., and Schimke, R. T., 1978, *J. Biol. Chem.* **253:**2483–2495.

Wozney, J., Hanahan, D., Morimoto, R., Boedtker, H., and Doty, P., 1981, *Proc. Natl. Acad. Sci. U.S.A.* **78:**712–716.

11

Transcription of Ribosomal RNA Genes in Mouse and Frog

BARBARA SOLLNER-WEBB, JoANNE KAYE
WILKINSON, KATHRYN G. MILLER,
RON WIDES, VAL CIZEWSKI,
RONALD REEDER, and JUDITH ROAN

1. INTRODUCTION

From information presented in this volume and elsewhere, it is now clear that major control of eukaryotic gene expression is exerted at the transcriptional level, apparently largely at the initiation step. Consequently, much interest is being devoted to analyzing the transcriptional process, both *in vivo* and *in vitro*. Studies in manipulable systems, especially those that utilize cloned DNA templates, are proving most fruitful. Systems in which RNA polymerase III mediates the accurate transcription of cloned 5 S RNA, VA RNA, and transfer RNA (tRNA) genes (Brown and Gurdon, 1977, 1978; Wu, 1978; Birkenmeier *et al.*, 1978; Weil *et al.*, 1979a) and systems in which RNA polymerase II accurately transcribes messenger RNA-coding genes (Weil *et*

BARBARA SOLLNER-WEBB, JoANNE KAYE WILKINSON, KATHRYN G. MILLER, RON WIDES, and VAL CIZEWSKI ● The Johns Hopkins University, School of Medicine, Baltimore, Maryland 21205.   RONALD REEDER and JUDITH ROAN ● Hutchinson Cancer Research Center, Seattle, Washington 98104.

al., 1979b; Mulligan et al., 1979; Hamer and Leder, 1979; Grosschedl and Birnsteil, 1980a; Manley et al., 1980; McKnight and Gavis, 1980) have been developed in the last five years. From studies with these systems, many aspects of the nucleoprotein interactions that direct eukaryotic transcription—that is, the nucleotide sequences and the protein species that facilitate and regulate this process—are being elucidated (Sakonju et al., 1980; Bogenhagen et al., 1980; Grosschedl and Birnsteil, 1980b; Fowlkes and Shenk, 1980; Engelke et al., 1980; Honda and Roeder, 1980; McKnight et al., 1981; Hu and Manley, 1981; Wasylyk and Chambon, 1981; Hofstetter et al., 1981; Segall et al., 1981; Matsuei et al., 1981; McKnight and Kingsbury, 1982).

In contrast to the study of genes transcribed by RNA polymerases II and III, the study of transcription of the ribosomal RNA (rRNA)* genes by RNA polymerase I has significantly lagged. However, accurate initation of cloned rDNA has recently been demonstrated in systems from mouse (K. Miller and Sollner-Webb, 1981; Grummt, 1981b; Mishima et al., 1981), frog (Sollner-Webb and McKnight, 1982; Wilkinson and Sollner-Webb, 1982), *Drosophila* (Krohne and Rae, 1982), and human (Miesfield and Arnheim, 1982) cells, and significant progress on the study of rDNA transcription is now being made. In this chapter, we summarize the transcription systems that we have developed for *Xenopus* and for mouse rRNA genes and the results that we have obtained concerning the mechanism of RNA polymerase I transcription.

2. *XENOPUS* RIBOSOMAL DNA TRANSCRIPTION

The *Xenopus laevis* rRNA genes, the first eukaryotic genes to be purified, have been the subject of extensive study and many reviews (Reeder, 1974; Fedoroff, 1979; Long and Dawid, 1980; Sollner-Webb et al., 1982). These genes are organized as tandem 40 S rRNA-coding regions alternating with spacer regions that are generally not transcribed. The transcription initiation site has been precisely located within the rDNA repeat at a position designated nucleotide + 1 (Sollner-Webb and Reeder, 1979). This conclusion is based

---

* Throughout this chapter, *rRNA* will refer to the large precursor RNA for the 18 S, 28 S, and 5.8 S RNAs of the ribosome, or portions thereof. In *Xenopus*, the "40 S" primary transcript is approximately 7.5 kilobases (kb); in mouse, the "45 S" primary transcript in approximately 14 kb. The abbreviation *rDNA* will refer to the DNA that codes for rRNA and to the spacer that separates adjacent rRNA-coding regions, or portions thereof.

on the fact that isolated 40 S rRNA from oocytes is polyphosphorylated, hence apparently has an unprocessed 5' end (Reeder *et al.*, 1977); the 5' sequence of this polyphosphorylated RNA defines nucleotide + 1. Until very recently, *X. laevis* was the only vertebrate the rDNA initiation site of which was precisely determined, so this species offered the unique possibility of developing an rDNA transcription system that could be verified to initiate faithfully. However, many early attempts to transcribe *Xenopus* rRNA genes *in vitro* were unsuccessful (reviewed in Sollner-Webb *et al.*, 1982). Purified polymerase has not yet been shown to initiate accurately, while with crude systems it is a major problem to distinguish the transcription under study from that inherent in the cell extract.

2.1. Transcription Systems

Now, cloned rDNA can be transcribed in a number of systems. Trendelenburg and Gurdon (1978) made a significant advance with the introduction of an electron-microscopic assay (O. Miller and Beatty, 1969) to examine transcription of cloned *X. laevis* rDNA microinjected into *X. laevis* oocytes. They found that a few of the circular, microinjected rDNA molecules contained transcription complexes of approximately the expected length, suggesting that initiation and termination are working correctly on the cloned rDNA, although at a very low level. Bakken *et al.* (1982) modified and improved this assay, demonstrating that this transcription is indeed by RNA polymerase I and that it functions with only a 430-base-pair (bp) region containing the *in vivo* rDNA initiation site. Unfortunately, this electron-microscopic method of analysis remains laborious and is not suitable for precisely locating the initiation site or for quantitative studies.

2.1.1. S1 Assay and Oocyte Microinjection

We thus have recently developed a simple biochemical assay for accurate initiation on cloned *X. laevis* rDNA (Sollner-Webb and McKnight, 1982). For this analysis, *X. laevis* rDNA is microinjected into *X. borealis* oocyte nuclei. The resultant *X. laevis* rRNA is then detected by S1 nuclease mapping using an *X. laevis*-specific hybridization probe (Fig. 1A). This probe is a 5'-end-labeled fragment that overlaps the rRNA initiation site, so the precise

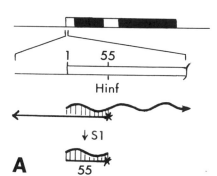

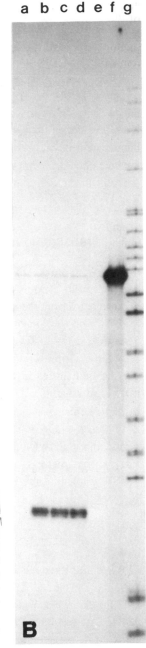

FIGURE 1.   S1 nuclease analysis of transcription on cloned rDNA. (A) Initiation region of the 40 S rRNA gene of *X. laevis*. A DNA probe, 5′-end-labeled at nucleotide 55 (*), is hybridized to the transcribed rRNA (∼) and then treated with S1 nuclease. (B) The S1 product is denatured, sized by polyacrylamide gel electrophoresis, and detected by autoradiography. Analysis of RNA from two *X. borealis* oocytes (a), one *X. laevis* oocyte (b), two *X. borealis* oocytes injected with cloned *X. laevis* rDNA (c), two *X. borealis* oocytes injected with cloned *X. laevis* rDNA and 1 mg/ml α-amanitin (d), tRNA (e), and untreated probe (f). (g) *Hpa*II-cleaved pBR322 marker.

location of the initiation site may be directly determined from the length of the S1-nuclease-resistant hybrid. Transcription on the cloned rDNA is thus seen to initiate faithfully (Fig. 1B, lanes b and c) and is catalyzed by RNA polymerase I (Fig. 1B, lane d). The endogenous *X. borealis* rRNA does not interfere with this assay, for it does not hybridize with the probe (Fig 1B, lane a). We have varied a number of parameters of the microinjection: the accuracy and extent of initiation are not highly dependent on the injection volume or injection buffer, but a closed circular rDNA template is required and the rDNA must be delivered to the nucleus. Up to one third as much *X. laevis* rRNA is made in the microinjected oocyte as is endogenously present

FIGURE 2. S1 nuclease analysis of *in vitro* transcription. The oocyte extract was centrifuged to yield a supernatant and a pellet, and fractions were assayed for their ability to transcribe cloned *X. laevis* rDNA. S1 assays of RNA from 0.5 *X. laevis* oocytes (a), reaction with complete oocyte extract (b), reaction with oocyte extract supernatent (c), reaction with oocyte extract pellet (d), and reaction with the pellet resuspended in the supernatant (e). (f) Sizing markers.

in an uninjected *X. laevis* oocyte; one injected cell's worth of RNA is ample for detection.

2.1.2. *In Vitro* Transcription

With this same S1 assay, we have also demonstrated accurate initiation on *X. laevis* rDNA *in vitro* (Fig. 2, lanes a and b). Cloned rDNA template is transcribed in a homogenate of manually isolated *X. borealis* oocyte nuclei (Wilkinson and Sollner-Webb, 1982). One major advantage of this cell-free system (in addition to its ease relative to oocyte microinjection) is that the

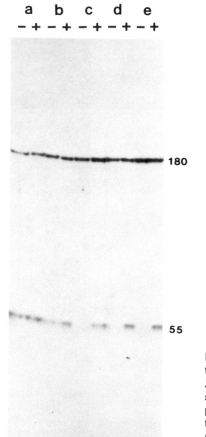

FIGURE 3. Dinucleotide stimulation of *in vitro* transcription. Various concentrations of cloned *X. laevis* rDNA template were added *in vitro* transcription reactions, unsupplemented (−) or supplemented (+) with ApG using the following template concentrations (μg/ml): (a) 5, (b) 10, (c) 20, (d) 35, (e) 50.

necessary components can be fractionated. Figure 2 illustrates the separation of the extract into two fractions, both of which are inactive alone (Fig. 2, lanes c and d), but reconstitute into a fully active extract (Fig. 2, lane e).

An interesting feature of the *in vitro* transcription system is its stimulation by dinucleotides. *Xenopus laevis* rDNA initiates with pppApG . . . (Sollner-Webb and Reeder, 1979), and addition of the dinucleotide ApG to the *in vitro* system can markedly increase the amount of specific transcription (Wilkinson and Sollner-Webb, 1982) (Fig. 3). The dinucleotides are acting as sequence-specific primers, for ApG is unique in its ability to stimulate accurate transcription, while GpG slightly stimulates production of a transcript starting at nucleotide $+2$ with GpG. Unlike the prokaryotic case (Glaser and Cashel, 1979), the mechanism of this dinucleotide stimulation is apparently not one of compensation for a limiting rXTP concentration. Rather, the stimulation occurs only at high template concentrations (Fig. 3). From analysis of the *in vitro* transcription products made under a variety of conditions, we conclude that (1) there are at least three components in the extract necessary for *in vitro* rRNA initiation and (2) the dinucleotide primer can augment or circumvent the need for one of these initiation factors (Wilkinson and Sollner-Webb, 1982).

2.1.3. Heterologous Transcription

Finally, we have a third way to transcribe *Xenopus* rDNA. This method uses a mouse tissue-culture cell extract, described below for its ability to transcribe mouse rDNA. With this heterologous system, two kinds of transcripts of *X. laevis* rDNA can be generated (Wilkinson and Sollner-Webb, 1982). Under low-salt conditions, accurate although inefficient initiation occurs. Alternatively, under the standard mouse transcription conditions, initiation is very efficient but occurs four nucleotides upstream from the site used *in vivo* (Fig. 4A). We note, however, that there is a strong (10/13 nucleotide) homology between the sequences about the *in vivo X. laevis* start site (nucleotide $+1$) and this alternate start site (Fig. 4B, nucleotide -4), suggesting that this sequence may be important in precise initiation-site selection. The success of this heterologous transcription also indicates that the RNA polymerase I reaction may not be as species-specific as previously reported (Grummt *et al.*, 1982).

2.2. Transcriptional Control Region

We are now using both the oocyte microinjection and the *in vitro* transcription system to identify the nucleotide sequences that specify rRNA initiation (Sollner-Webb *et al.*, 1983). To this end, we assay the transcriptional potential of deletion mutants constructed *in vitro* by exonuclease treatment of the rDNA (5' deletion mutants all have nucleotides upstream of the designated nucleotide deleted, while 3' deletion mutants all have nucleotides downstream of the designated nucleotide deleted). Our various transcription systems are complementary, for they require different nucleotide regions to

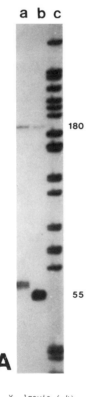

a b c

180

55

A

FIGURE 4. Transcription of *X. laevis* rDNA in the mouse cell-free extract. (A) S1 nuclease analysis of the product of the heterologous transcription (a) and of control *X. laevis* rRNA (b). (c) Sizing markers. (B) Sequence comparison of the initiation region of *X. laevis* rDNA and this same sequence displaced by four nucleotides. Heavy lines represent identities.

B

X. laevis (-4)   T T T G G C A T G T ▬▬ G ▬▬ C ▬▬▬ A ▬▬▬ T A G G G G A A G A C

X. laevis (+1)   G C A T G T G C G G G C A G G A A G G T A G G G G A A G A C C G G C

promote synthesis (Fig. 5A). We believe that these systems recognize different elements of the *in vivo* initiation signal.

We were surprised to find that in the microinjected oocyte, only a very small rDNA region is needed for initiation. The 5' deletions extending into nucleotide -7 (5' del -7) still initiate transcription, while faithful initiation has not been detected with 5' del $+2$ or more extensive 5' deletions. Experiments on the minimal 3' boundary are still in progress, but 3' del $+6$ can direct accurate initiation, while a 3' del -11 does not. Further evidence that the region immediately surrounding nucleotide $+1$ is necessary for initiation is derived from the observation that a mutant where the nucleotides between -9 and $+2$ are replaced by a *Bam* linker (6 of 10 positions altered) does not direct accurate initiation. Thus, the 5' boundary of the minimal region needed for accurate initiation is between nucleotides -7 and $+2$, and the 3' boundary appears to be between nucleotides $+6$ and -11. Certainly, this does not mean that other regions lack a functional role in specifying or controlling transcriptional initiation, only that the region between nucelotides -7 and $+6$ is essential to initiate rRNA in microinjected oocytes.

In contrast, a large region of 5' flanking sequence is involved in directing initiation in the frog *in vitro* system. All 5' deletions extending up to 5' del -142 initiates accurately and efficiently, while 5' del -127 and all more extensive deletions do not (Fig. 5B). Thus, approximately 140 more nucleotides of 5' flanking region are needed for initiation *in vitro* than in the microinjected oocyte. The 3' border of the region required for initiation *in vitro* is currently under investigation, but all 3' deletions extending up to 3' del $+16$ initiate accurately.

We believe that the two transcription systems are recognizing different aspects of a complex regulatory region that directs *Xenopus* rDNA initiation. Certainly, one major difference between the two systems is that the transcription factors are considerably more dilute in the oocyte extract than in the intact oocyte; under these conditions, a more extensive control region may be needed. (In accordance with this suggestion, in extracts from certain frogs that initiate very efficiently, small amounts of accurate initiation are obtained *in vitro* with 5' deletions between -127 and -7.) Such nested control regions have a precedent in RNA-polymerase-II-mediated transcription (see references above), but in that case the *in vitro* system can function with only a small DNA region, while the microinjected oocyte requires an extensive 5' flanking region, the reverse of the *Xenopus* RNA polymerase I case.

Using our third method of transcribing *Xenopus* rDNA, the mouse tissue-culture cell extract, the same large rDNA region is needed as is required for initiation in the *Xenopus in vitro* system. On the 5' side, the boundary is between 5' del − 142 and 5' del − 127; on the 3' side, the boundary is upstream of nucleotide + 6. The fact that the same nucleotide region specifies *in vitro* initiation in the homologous and the heterologous system indicates that the transcription in the mouse cell extract is responding to an authentic initiation signal and not to a fortuitous one. It is curious, however, that this heterologous initiation in the mouse cell extract requires such an extensive 5' flanking region, since only 35 nucleotides of 5' flanking region are required to transcribe the homologous mouse rDNA in this same extract (see below). It thus appears that the mouse and frog initiation signals are conserved—at

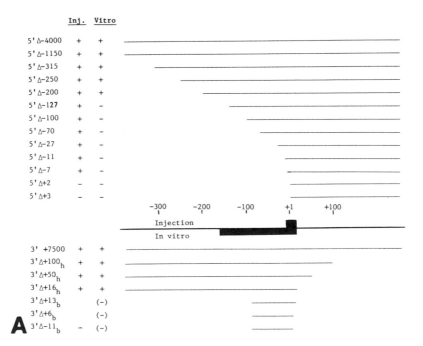

FIGURE 5. Analysis of 5' deletions of *X. laevis* rDNA. (A) Summary of the deletion data obtained with our two *Xenopus* transcription systems. Analysis of frog rDNA in the mouse cell extract gives the same results as the frog *in vitro* system. (B) Deletions of the rDNA extending toward the initiation site from the upstream side were assayed for their transcriptional ability using the *Xenopus in vitro* extract and S1 analysis. The first six lanes show

least to the extent that mouse cell factors can recognize and act on both species' rDNA—yet the sequences that direct initiation are apparently located at different positions relative to the rRNA start sites.

2.3. Initiation in the *Xenopus* Nontranscribed Spacer

While the major transcript of *Xenopus* rDNA is 40 S rRNA, occasionally transcription units are also observed in the region that separates the tandem

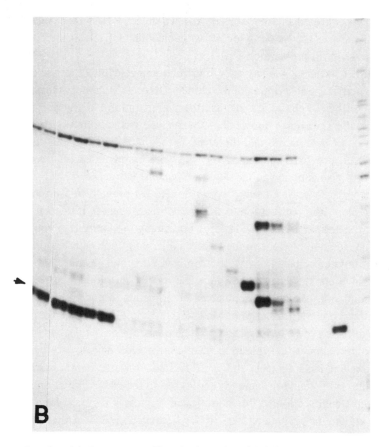

increasing size deletions approaching the boundary (5′ del: −1050, −250, −240, −210, −166, −158), while the next 11 lanes show increasing size deletions extending beyond the boundary (5′ del: −127, −115, −95, −85, −75, −65, −45, −27, −22, −11, −7). Next is a control mapping of *X. laevis* oocyte rRNA, and their final lane contains size markers.

rRNA genes, in the so-called "nontranscribed spacer" (Scheer *et al.*, 1973, 1977; Franke *et al.*, 1976, reviewed in Rungger and Crippa, 1977; Sollner-Webb *et al.*, 1982). These RNAs, termed *prelude transcripts,* appear to initiate at discrete sites (Trendelenburg, 1982). On sequencing of *X. laevis* rDNA, it was noted that the spacer contains regions that have approximately 90% sequence homology to the rRNA initiation site (nucleotides −145 to +4), and it was suggested that prelude transcription initiates at these sites (Bosely *et al.*, 1979; Sollner-Webb and Reeder, 1979; Moss and Birnstiel, 1979; Moss *et al.*, 1979, 1980). Using S1 mapping of total oocyte RNA, we have now confirmed this hypothesis, for we detect RNAs that begin at the site of the spacer that is homologous to nucleotide +1 (B. Sollner-Webb *et al.*, manuscript in preparation).

Further, when *X. laevis* rDNA is microinjected into the nuclei of *X. borealis* oocytes, transcription of the cloned rDNA also initiates at these spacer sites (B. Sollner-Webb *et al.*, manuscript in preparation). Using this microinjection system, we have shown that the prelude transcription is resistant to high levels of α-amanitin and therefore is mediated by RNA polymerase I, as had been previously assumed but not yet demonstrated. Further, these spacer sites also function in the *in vitro Xenopus* extract.

The size of the 5′ flanking region needed for prelude transcription provides an interesting correlation to the size of the 5′ region needed for rRNA initiation, determined above. Plasmids containing the entire region of spacer with sequence homologous to the rDNA initiation region (nucleotides −145 to +4) initiate prelude transcription both *in vitro* and in the microinjected oocyte. In contrast, plasmids that lack the first 20 nucleotides of this region (those containing the spacer sequences homologous to nucleotides −125 to +4) initiate prelude transcription on microinjection, but not *in vitro*. Thus, *in vitro* initiation of both rRNA and prelude transcription requires an extensive 5′ flanking region, while initiation in microinjected oocytes does not.

From the facts that (1) the prelude initiation site has preserved 145 nucleotides of 5′ flanking region homologous to the rRNA initiation site and (2) the upstream section of this homologous region is required for initiation *in vitro* at both the prelude and the rRNA start site, we deduce that an *in vivo* RNA polymerase I control region extends approximately 140 bp upstream from the actual initiation site. However, since accurate intiation occurs in the microinjected oocyte at both these sites in the absence of this extensive 5′ flanking region, information close to the initiation site can be sufficient to direct faithful transcription under appropriate conditions.

3. MOUSE RIBOSOMAL DNA TRANSCRIPTION

The rRNA genes of the mouse have also been extensively studied (reviewed in K. Miller and Sollner-Webb, 1982). Like those of *Xenopus* and apparently all other higher eukaryotes, the mouse rRNA genes are also organized as tandem head-to-tail repeats; in this case, there is a very large transcribed region [14 kilobases (kb)] that alternates with about 25 kb of spacer (Cory and Adams, 1977). The transcription initiation site is now known to reside 4.5 kb upstream from the 18 S rRNA coding region at the site illustrated in Fig. 6A (K. Miller and Sollner-Webb, 1981).

3.1. Transcription Systems

Prior to mapping of the initiation site, investigators were already attempting to develop *in vitro* transcription systems for mouse rDNA (reviewed in K. Miller and Sollner-Webb, 1982). In early studies, the synthesis was apparently due to elongation *in vitro* by RNA polymerase I molecules that had initiated *in vivo* or to aberrant *in vitro* initiation. Grummt (1981a) made a major step forward when she utilized purified RNA polymerase I and an

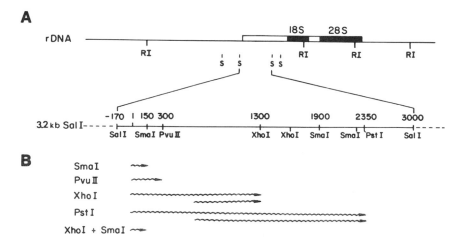

FIGURE 6. Mouse rDNA map. (A) Repeating unit of mouse rDNA (*top*) and a map of the 3.2-kb *Sal*I fragment surrounding the initiation site (*bottom*). (B) Runoff RNAs produced from rDNA templates truncated with the indicated enzymes.

S100 cell extract to effect specific synthesis on cloned mouse rDNA; however, her original mapping of the initiation site was in error. We then began working with a somewhat different S100 extract of mouse tissue-culture cells and found that the *in vitro* initiation site is actually 450 nucleotides upstream from the site identified by Grummt (K. Miller and Sollner-Webb, 1981). This determination was made by runoff transcription analysis, using templates truncated at various positions (Fig. 6B), and was confirmed by S1 nuclease analysis (Fig. 7). From our mapping, we immediately conclude that initiation can occur without an extensive region of 5′ flanking rDNA, since the rDNA cloning site is at nucleotide − 170; further, the template may be cleaved at this site without impairing initiation. Thus, the region around nucleotide + 1 must contain all the information needed for accurate initiation (K. Miller and Sollner-Webb, 1981).

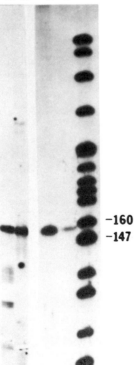

FIGURE 7.   Identity of *in vivo* and *in vitro* 5′ ends. Mouse rRNA made *in vitro* was subjected to S1 nuclease mapping using probe labeled at the *Sma*I site, nucleotide + 150 (lane c). The protected fragment comigrates with that protected by *in vivo* RNA (lanes a and d), which in turn comigrates with runoff RNA transcribed *in vitro* from *Sma*I-cleaved template (lane b). (e) Size markers, Hpa II-digested pBR322.

3.2.   Initiation Site

It is now clear that nucleotide $+1$ is precisely the initiation site both *in vitro* and *in vivo*. For the *in vitro* analysis (Wilkinson *et al.*, 1983), the 5' end of mouse rRNA was first mapped on the rDNA sequence by electrophoresing an S1 nuclease mapping analysis of mouse rRNA in the lane adjoining a chemical sequence analysis of the probe DNA; this located the presumptive

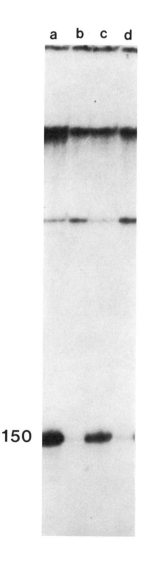

FIGURE 8.   Identification of initiation site by the use of dinucleotide primers. *In vitro* runoff transcription products made with the following nucleotides and dinucleotides: (a) 500 μM rXTP, no dinucleotide; (b) 30 μM rXTP, no dinucleotide; (c) 30 μM rXTP, 300 μM ApC; (d) 30 μM rXTP, 200 μM ApG.

initiation site at the A residue within a GGTACT sequence (nucleotides −3 to +3). To confirm that this site is indeed the *in vitro* initiation site, we used dinucleotide analysis (Fig. 8). The rXTP concentration in the transcription reaction was dropped from the usual 500 μM (lane a) to 30 μM (lane b) to inhibit initiation, and various dinucleotides were added to this inhibited re-

FIGURE 9. Mapping of the 5′ boundary of mouse rDNA needed for initiation. (A) The first five lanes show runoff RNAs transcribed from 5′ deletions of increasing size, ranging from 5′ del −127 to 5′ del −35; the next four lanes show RNAs from deletions of increasing size between 5′ del −26 and 5′ del +25. The last two lanes show RNA made from the parental plasmid (5′ del −170) and a DNA sizing standard. A plasmid that has more than 4 kb of 5′ flanking region transcribes no more effeciently than does 5′ del −170. (B) Summary of 5′ and 3′ deletion mapping data for mouse rDNA and diagram of the boundaries of the essential region in relation to the region needed for frog rRNA initiation.

action. Only ApC was able to stimulate correct transcription (lane c). From various lines of evidence, we are convinced that ApC is acting as a site-specific and sequence-specific primer to circumvent the initiation block caused by the limiting substrate. This, in turn, indicates that mouse rRNA starts with ApC . . . *in vitro*, i.e., that it starts at nucleotide +1. The largest detected rRNA isolated from mouse cells also maps at nucleotide +1 by S1 nuclease analysis (K. Miller and Sollner-Webb, 1981). Further, this *in vivo* RNA has been shown to bear a 5' polyphosphate group at precisely nucleotide +1 (Grummt, 1981b), indicating that nucleotide +1 is actually the *de novo* initiation site *in vivo* as it is *in vitro*.

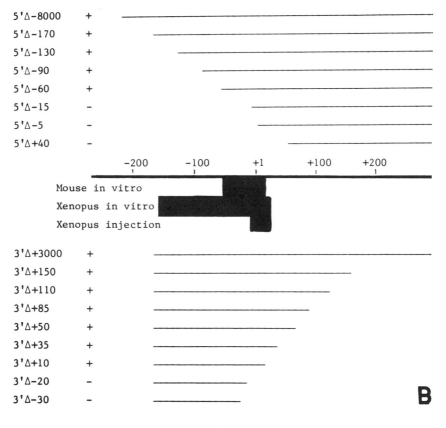

FIGURE 9. *Continued*

3.3. New Ribosomal RNA Processing Site

We have detected an rRNA processing site 650 nucleotides downstream from the initiation site (K. Miller and Sollner-Webb, 1981). It was first observed that 5' ends of rRNA synthesized *in vitro* in fact map to two sites on the rDNA, nucleotide +1 and nucleotide 650. Using pulse–chase and other experiments, we have demonstrated that the species with its 5' end at nucleotide 650 is a processing product of the species that initiated at nucleotide +1. The site at nucleotide 650 corresponds to the 5' end of the major species of 45 S rRNA isolated from cells (Urano *et al.*, 1980; K. Miller and Sollner-Webb, 1981), indicating that this processing occurs *in vivo* as well. This kind of an rRNA processing event, within the external transcribed spacer, was unanticipated, since all rRNA processing sites previously mapped were at the ends of the mature rRNA species. There is recent evidence that this type of processing also occurs in human rRNA maturation (Financsek *et al.*, 1982).

3.4. Transcription Control Region

We are currently studying the nucleotide sequences and the protein components that function to specify initiation of mouse rDNA. The necessary nucleotide sequences are being mapped by the *in vitro* genetic approach in which 5' and 3' deletions of the cloned rDNA template are constructed and assayed for their ability to initiate transcription (K. Miller and B. Sollner-Webb, manuscript in preparation). For mouse rDNA assayed in the *in vitro* system, 5' del −35 and less extensive deletions initiate runoff transcription accurately and quantitatively, while 5' del −26 and more extensive deletions fail to initiate (Fig. 9A). On the 3' side, 3' del +12 initiates, while 3' del −10 does not. Hence, the borders of the region necessary for mouse rRNA initiation reside between nucleotides −35 and +12 (Fig. 9B).

Curiously, the transcriptional potential of the 3' deletions is markedly affected by the construction of the template plasmid. The 3' deletions cloned clockwise into the *Pvu*II site of pBR322 apparently produce less rRNA than do control plasmids or 3' deletions cloned clockwise into the *Hind*III site of pBR322.

3.5. Stable Transcription Complex

A somewhat different set of experiments has demonstrated that mouse rDNA transcription occurs via the formation of a stably activated template. As one might anticipate, when two runoff templates of different lengths are added to the transcription reaction simultaneously, both runoff RNAs are produced, Figure 10a (lanes a, b, and e) shows this result with templates linearized at the *Sma*I site ($+150$) and at the *Pvu*II site ($+300$). However, when this experiment is modified so that one template is added 15 min before the second (and labeling starts after this second addition), RNA is transcribed only from the first template (Fig. 10B, lanes c and d). Apparently, a factor necessary for transcription is titrated out on the first template. Further, this complex is stable, for the factor does not transfer to the second template to permit its transcription, even after over an hour of continued incubation (Cizewski and Sollner-Webb, 1983).

To purify the transcription complex from the rest of the mouse cell extract, we employed centrifugation. Since the S100 extract is the soluble component of a 100,000g centrifugation of lysed cells, it is not surprising that almost all the protein of the transcription reaction remains in the super-

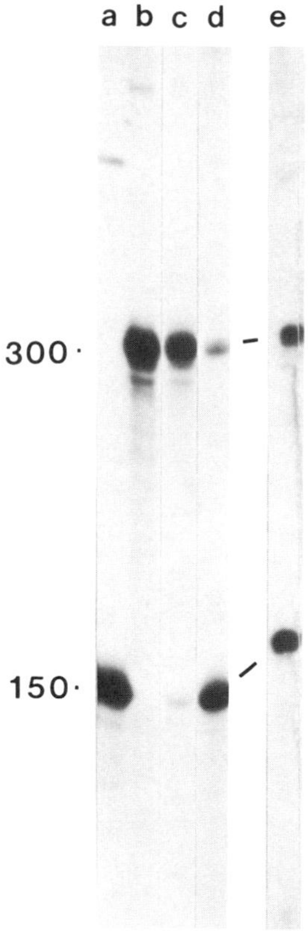

FIGURE 10. Sequential template addition experiments. Template cleaved with *Sma*I ($+150$) and template cleaved wtih *Pvu*II ($+300$) were used for *in vitro* transcription. Lanes: (a) Only *Sma*I-cleaved template; (b) only *Pvu*II-cleaved template; (c) *Pvu*II-cleaved template added at time 0 and *Sma*I-cleaved template and label added after 15 min; (d) *Sma*I-cleaved template added at time 0 and *Pvu*II-cleaved template and label added after 15 min; (e) *Sma*I- and *Pvu*II-cleaved template added simultaneously.

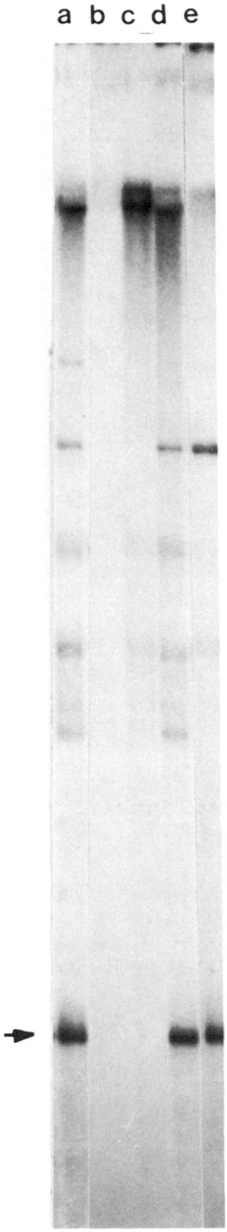

a b c d e

FIGURE 11.   A transcription complex that lacks polymerase I.
Runoff transcription products from a complete reaction (a); a
reaction preincubated for 15 min, centrifuged, the supernatant
discarded, and the pellet resuspended in transcription buffer
(b); the supernatant from lane b (c); a pellet resuspended in the
supernatant (d); and a complete reaction (e). Labeling starts
after the final resuspension.

natant of a 10,000*g* spin; most of the added template also remains in the soluble fraction. Fortuitously, a component necessary for transcription partitions cleanly into the pellet while another remains in the supernatant (Fig. 11). That is, neither the supernatant nor the resuspended pellet is able to transcribe accurately, even in the presence of additional template (lanes b and c). Yet the necessary components were not destroyed on centrifugation, because the two fractions reconstitute into a fully active system (lane d). By combining the supernatants and the pellets made with different templates, it is straightforward to show that the active template partitioned to the pellet: a pellet from a *Sma*I (+150)-cleaved template plus the supernatant from a *Pvu*II (+300)-cleaved template generate a 150-nucleotide RNA (and vice versa) (V. Cizewski, R. Wides, and B. Sollner-Webb, manuscript in preparation).

The identity of the important component of the supernatant was suggested by the observation that more than 95% of the RNA polymerase I activity remained in this fraction. To test the hypothesis that a polymerase-associated protein is all that the supernatant contributes, we added partially purified RNA polymerase I to the pellet fraction and found that this addition resulted in copious amounts of correct transcription (B. Sollner-Webb, K. Rose, and S. Jacobs, unpublished observations). Each activated template in the pellet is responsible for the synthesis of about 20 RNA molecules. From these data, we draw the following model for mouse rDNA transcription *in vitro:* Synthesis requires rDNA and RNA polymerase I as well as other factors. Some of these other factors associate with specific sequences in the rDNA to form stable transcription complexes that persist on the DNA without exchange through multiple rounds of initiation. When these stable complexes are isolated and suplemented with polymerase-associated proteins, either partially purified or as a crude supernatant, active transcription ensues.

4. SUMMARY

We have studied transcription of cloned mouse and frog rDNA in a number of systems. In the mouse extract, mouse rDNA requires a 5′ flanking region of less than 40 bp to initiate. Using this same mouse extract, however, initiation of *Xenopus* rDNA needs a 5′ flanking region of approximately 140 bp. *Xenopus* rDNA also requires this approximate-140-bp region to initiate transcription in the *Xenopus* cell-free extract. The fact that the same *Xenopus* rDNA region is recognized in both *in vitro* systems suggests that the *Xenopus* initiation signal is recognized by a factor conserved in evolution. Further, initiation of *Xenopus* rDNA in the microinjected oocyte utilizes yet a different sequence; only seven nucleotides of 5′ flanking region are sufficient to obtain correct initiation in this case. We conclude that each of our transcription systems highlights different aspects of the interactions that control rRNA synthesis *in vivo*.

In all our *in vitro* transcription systems, synthesis occurs on templates that are stably associated with activating proteins. Although these complexes are inactive when isolated, addition of the remaining extract components allows synthesis to occur. We are currently investigating what components are involved in the formation of these transcription complexes and their role in the initiation of rRNA synthesis.

REFERENCES

Bakken, A., Morgan, G., Sollner-Webb, B., Roan, S., Busby, S., and Reeder, R., 1982, *Proc. Natl. Acad. Sci. U.S.A.* **79**:56.

Birkenmeier, E. Brown, D., and Jordan, E., 1978, *Cell* **15**:1075.

Bogenhagen, D., Sakonju, S., and Brown, D., 1980, *Cell* **19**:27.

Boseley, P., Moss, T., Machler, M., Portmann, R., and Birnstiel, M., 1979, *Cell* **17**:19.

Brown, D., and Gurdon, J., 1977, *Proc. Natl. Acad. Sci. U.S.A.* **74**:2064.

Brown, D., and Gurdon, J., 1978, *Proc. Natl. Acad. Sci. U.S.A.* **75**:2849.

Cizewski, V., and Sollner-Webb, B., 1983, *Nucleic Acids Res.* (in press).

Cory, S., and Adams, J., 1977, *Cell* **11**:795.

Engelke, D., Ng, S., Shastry, B., and Roeder, R., 1980, *Cell* **19**:717.

Fedoroff, N., 1979, *Cell* **16**:697.

Financsek, I., Mizumoto, K., Mishima, Y., and Muramatsu, M., 1982, *Proc. Natl. Acad. Sci. U.S.A.* **79**:3092.

Fowlkes, D., and Shenk, T., 1980, *Cell* **22**:405.

Franke, W., Scheer, U., Spring, H., Trendelenburg, M., and Krohne, G., 1976, *Exp. Cell Res.* **100**:233.

Glaser, M., and Cashel, M., 1979, *Cell* **16**:111.

Grosschedl, R., and Birnsteil, M., 1980a, *Proc. Natl. Acad. Sci. U.S.A.* **77**:1432.

Grosschedl, R., and Birnstiel, M., 1980b, *Proc. Natl. Acad. Sci. U.S.A.* **77**:7102.

Grummt, I., 1981a, *Proc. Natl. Acad. Sci. U.S.A.* **78**:727.

Grummt, I., 1981b, *Nucleic Acids Res.* **9**:6091.

Grummt, I., Roth, E., and Paule, M., 1982, *Nature (London)* **296**:173.

Hamer, D., and Leder, P., 1979, *Nature (London)* **281**:35.

Hofstetter, H., Kressman, A., and Birnstiel, M., 1981, *Cell* **24**:573.

Honda, B., and Roeder, R., 1980, *Cell* **22**:119.

Hu, S., and Manley, J., 1981, *Proc. Natl. Acad. Sci. U.S.A.* **78**:820.

Krohne, B., and Rae, P., 1982, *Proc. Natl. Acad. Sci. U.S.A.* **79**:1501.

Long, E., and Dawid, I., 1980, *Annu. Rev. Biochem.* **49**:727.

Manley, J., Fire, A., Cano, A., Sharp, P., and Gefter, M., 1980, *Proc. Natl. Acad. Sci. U.S.A.* **77**:3855.

Matsuei, T., Segall, J., Weil, P., and Roeder, R., 1981, *J. Biol. Chem.* **255**:11,992.
McKnight, S., and Gavis, E., 1980, *Nucleic Acids Res.* **8**:5931.
McKnight, S., and Kingsbury, R., 1982, *Science* **217**:180.
McKnight, S., Gavis, E., Kingsbury, R., and Axel, R., 1981, *Cell* **25**:385.
Miesfeld, R., and Arnheim, N., 1982, *Nucleic Acids Res.* **10**:3933.
Miller, O., and Beatty, B., 1969, *Science* **164**:955.
Miller, K., and Sollner-Webb, B., 1981, *Cell* **27**:165.
Miller, K., and Sollner-Webb, B., 1982, in: *The Cell Nucleus* (H. Busch, ed.), Vol. 12, Academic Press, New York, p. 69.
Mishima, Y., Yamamoto, O., Kominami, R., and Muramatsu, M., 1981, *Nucleic Acids Res.* **9**:6773.
Moss, T., and Birnsteil, M., 1979, *Nucleic Acids Res.* **6**:3733.
Moss, T., Boseley, P., and Birnsteil, M., 1980, *Nucleic Acids Res.* **8**:467.
Mulligan, R., Howard, B., and Berg, P., 1979, *Nature (London)* **277**:108.
Reeder, R., 1974, in: *Ribosomes* (M. Nomura *et al.*, eds.), Cold Spring Harbor Laboratory, Cold Spring Harbor, New York, p. 489.
Reeder, R., Sollner-Webb, B., and Wahn, H., 1977, *Proc. Natl. Acad. Sci. U.S.A.* **74**:5402.
Rungger, D., and Crippa, M., 1977, *Prog. Biophys. Mol. Biol.* **31**:247.
Sakonju, S., Bogenhagen, D., and Brown, D., 1980, *Cell* **19**:13.
Scheer, U., Trendelenburg, M., Krohne, G., and Franke, W., 1973, *Exp. Cell Res.* **80**:175.
Scheer, U., Trendelenburg, M., and Franke, W., 1977, *Chromosoma* **60**:147.
Segall, S., Matsui, T., and Roeder, R., 1981, *J. Biol. Chem.* **255**:11,986.
Sollner-Webb, B., and McKnight, S., 1982, *Nucleic Acids Res.* **10**:3391.
Sollner-Webb, B., and Reeder, R., 1979, *Cell* **19**:485.
Sollner-Webb, B., Wilkinson, J., and Miller, K., 1982, in: *The Cell Nucleus* (H. Busch, ed.), Vol. 12, Academic Press, New York, p. 31.
Sollner-Webb, B., Wilkinson, J., Roan, J., and Reeder, R., 1983, *Cell* **35**:199.
Trendelenburg, M., 1982, *Biol. Cell* **42**:1.
Trendelenburg, M., and Gurdon, J., 1978, *Nature (London)* **276**:292.
Urano, Y., Kominami, R., Mishima, Y., and Muramatsu, M., 1980, *Nucleic Acids Res.* **8**:6043.
Wasylyk, B., and Chambon, P., 1981, *Nucleic Acids Res.* **9**:1813.
Weil, P., Luse, D., Segall, J., and Roeder, R., 1979a, *Cell* **18**:469.
Weil, P., Segall, J., Harris, B., Ng, S.-Y., and Roeder, R., 1979b, *J. Biol. Chem.* **254**:6163.
Wilkinson, J., and Sollner-Webb, B., 1982, *J. Biol. Chem.* **257**:14375.
Wilkinson, J., Miller, K., and Sollner-Webb, B., 1983, *J. Biol. Chem.* (in press).
Wu, G., 1978, *Proc. Natl. Acad. Sci.* **75**:2175.

12

Nucleotide Sequence and Structure Determination of Rabbit 18 S Ribosomal RNA

JOHN F. CONNAUGHTON, AJIT KUMAR, and RAYMOND E. LOCKARD

1. INTRODUCTION

Ribosomes accomplish their translational role within a highly organized nucleoprotein structure. The specific aim for research on ribosomes, therefore, is to elucidate the architecture and eventually the function of these multimolecular complexes. Prokaryotic 70 S ribosomes can be dissociated into a small 30 S subunit and a large 50 S subunit. The 30 S subunit contains 16 S ribosomal RNA (rRNA) in association with 21 proteins, while the 50 S subunit contains a 5 S rRNA and the 23 S rRNA combined with 31 proteins. Eukaryotic 80 S ribosomes are appreciably larger. Their small 40 S subunit is comprised of 18 S rRNA and approximately 30 proteins, while the large 60 S subunit is much more complex, with one molecule each of 5 S, 5.8 S, and 28 S rRNA and 45–50 proteins.

In both prokaryotes and eukaryotes, the small subunit is a focal point where specific protein factors, initiator met-transfer RNA (tRNA), and messenger RNA (mRNA) organize into a functional initiation complex during protein synthesis. In recent years, considerable data on the structural and

JOHN F. CONNAUGHTON, AJIT KUMAR, and RAYMOND E. LOCKARD ● Department of Biochemistry, The George Washington University School of Medicine, Washington, D.C. 20037.

functional organization of the prokaryotic 16 S rRNA have been generated (Noller and Woese, 1981; Lake, 1980). Research on prokaryotic ribosomes has generally outpaced similar studies in eukaryotes, due in part to the ability to generate uniformly ^{32}P-labeled prokaryotic rRNA of high specific activity. The only complete nucleotide sequences available for eukaryotic 18 S rRNA have been deduced from DNA sequences of both the yeast and *Xenopus laevis* ribosomal genes (Rubstov *et al.*, 1980; Salim and Maden, 1981). A complete nucleotide sequence for a mammalian 18 S rRNA is not yet available. Furthermore, secondary structure predictions for eukaryotic 18 S rRNA are at present crude and are based entirely on extrapolations from the proposed prokaryotic model by searching for compensating base changes in the yeast and *Xenopus* sequences (Zwieb *et al.*, 1981; Stiegler *et al.*, 1981a).

In this chapter, we report the primary structure of nearly 70% of rabbit 18 S ribosomal RNA by sequence analysis of the rRNA directly. Highly purified rabbit 18 S rRNA was labeled *in vitro* at either its 5' or 3' terminus with ^{32}P, and nucleotide sequences were determined enzymatically and chemically by rapid sequencing gel analysis (Donis-Keller *et al.*, 1977; Lockard *et al.*, 1978; Peattie, 1979). Internal regions were sequenced from specific fragments generated by partial T_1 RNase digestion. We also report the first direct structure analysis of both the 5' and 3' domains of a eukaryotic 18 S rRNA using structure-specific enzymatic probes in solution.

2. METHODS

The methods for purification, ^{32}P-end-labeling, and chemical and enzymatic sequence analysis of 18 S rRNA are the same as previously described (Lockard *et al.*, 1982). Methods for structure-mapping ^{32}P-end-labeled 18 S rRNA using single- and double-strand-specific enzymatic probes are identical to those previously detailed (Lockard and Kumar, 1981). Fragments of 18 S rRNA were prepared by digestion with 2.5×10^{-5} U T_1ribonuclease (RNase)/μg RNA in structure buffer (20 mM Tris-HCl, pH 7.5, 300 mM KCl, 20 mM MgCl$_2$) for 15 min at 4°C. The reaction was terminated by incubation with 300 μg/ml of autodigested proteinase K for 25 min at 25°C. Following phenol extraction and ethanol precipitation, the T_1-RNase-generated fragments were 5'-^{32}P-end-labeled with [γ-^{32}P]-ATP and T_4 polynucleotide kinase and fractionated on an 80-cm-long 3.5% polyacrylamide gel run in 7 M urea. The 5'-^{32}P-end-labeled T_1 fragments were excised from the preparative gel and sequenced as previously detailed (Lockard *et al.*, 1982).

3. RESULTS

3.1. Nucleotide Sequence Determination

Initial nucleotide sequence data for rabbit 18 S ribosomal RNA were obtained using highly purified ^{32}P-end-labeled rRNA prepared as outlined in Fig. 1. Both the 5'- and 3'-^{32}P-end-labeled 18 S rRNAs can be easily purified on 3.5% acrylamide gels as shown in Fig. 2, then recovered after autora-

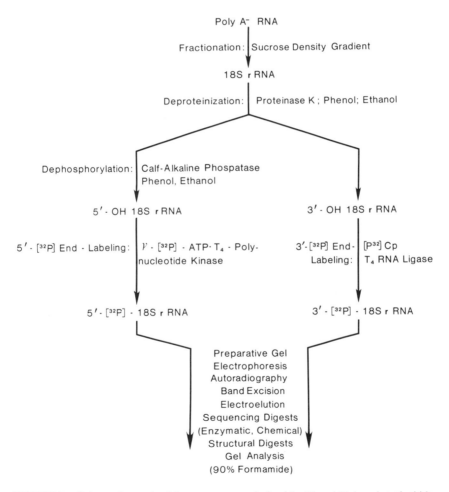

FIGURE 1. Scheme for nucleotide sequence analysis of the 5' and 3' domains of rabbit 18 S rRNA.

5′– [32P] · 18S r RNA   3′– [32P] · 18S r RNA

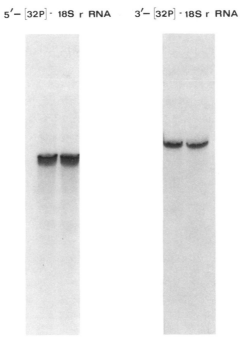

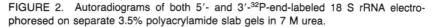

FIGURE 2.   Autoradiograms of both 5′- and 3′-³²P-end-labeled 18 S rRNA electrophoresed on separate 3.5% polyacrylamide slab gels in 7 M urea.

diography by excision of the gel piece and electroelution of the radioactive RNA into a dialysis bag. End-group analysis of the 5′-³²P-end-labeled 18 S rRNA revealed greater than 95% [³²P]-U after complete digestion with nuclease P_1 and analysis by thin-layer chromatography (Lockard and Raj-Bhandary, 1976), while analysis of the 3′-³²P-end-labeled 18 S rRNA after a limit digest with T_2 RNase revealed greater than 90% A-[³²P].

Nucleotide sequence analysis of the ³²P-end-labeled 18 S rRNAs was achieved by both base-specific enzymatic and chemical partial digestions according to previously published protocols (Donis-Keller *et al.*, 1977; Lockard *et al.*, 1978, 1982; Peattie, 1979). The A > G and G chemical reactions were performed on 5′-³²P-end-labeled 18 S rRNA, and both the purine and the pyrimidine chemical reaction with 3′-³²P-end-labeled RNA. Figure 3 shows an autoradiogram of a typical 10% polyacrylamide sequencing gel indicating the nucleotide sequence of residues 235–400 from the 5′-³²P-end-labeled terminus. Sequencing gels polymerized in 90% formamide rather than

8 M urea are much more effective in completely denaturing fragments in highly structured regions of the rRNA molecule. More than 150 nucleotides can usually be read off on a single 140-cm-long gel. Using just intact ^{32}P-end-labeled RNA, we were able to determine the 5′ proximal 400 residues and the 3′ distal 301 nucleotides for rabbit 18 S rRNA.

To extend the sequence analysis of 18 S rRNA to internal regions, large fragments of the RNA were generated by partial digestion with T_1 RNase, labeled at their 5′ termini with ^{32}P, and fractionated on an 80-cm-long preparative 3.5% polyacrylamide gel as shown in Fig. 4. Partial digestion with T_1 RNase in high-salt buffer (300 mM KCl, 20 mM $MgCl_2$, 20 mM Tris-HCl, pH 7.5) reproducibly generated several major fragments. Fragments 1, 2, and 3 were of sufficient purity and high enough specific activity to extend the sequence from the 3′ terminus to 804 nuclcotides.

Figures 5A and B summarize the nucleotide sequences at the 5′ and 3′ domains, respectively. Nucleotides designated either as X or Y* are likely modified, as determined from our base-specific enzymatic and chemical digestion patterns. The modified nucleotides that have already been identified in T_1 RNase fragments of Novikoff hepatoma (Choi and Busch, 1978), HeLa (Maden and Salim, 1974), and *X. laevis* (Maden and Kahn, 1977), having an identical sequence, are indicated parenthetically below. The cleavage sites generated with T_1 RNase for fragments 1, 2, and 3 begin at nucleotide residues 803, 711, and 487, respectively, as shown in Fig. 5B.

A phylogenetic comparison of the 1204 residues determined thus far for rabbit 18 S rRNA with both *X. laevis* and yeast nucleotide sequences is shown in Fig. 6. The pattern of sequence homology among the mammalian, frog, and yeast 18 S rRNAs shows conserved regions that are interspersed with tracts having little or no homology. The most striking differences can be found within the 5′ domain shown in Fig. 6A. The rabbit 5′ sequence compared with yeast shows seven major clusters of putative insertions (U_{120}–G_{123}; C_{179}, G_{180}, G_{187}, G_{201}, C_{202}, U_{207}, C_{209}; C_{235}–C_{238}; A_{242}–U_{244}, C_{248}–C_{252}; G_{259}–G_{272}; G_{280}, C_{281}; U_{297}, G_{309}, A_{310}, A_{315}, Y_{316}), resulting in a 33-nucleotide increase in the size of this region compared with yeast. Most of these insertions as well as the 75 base differences between the rabbit and yeast sequences result in a remarkable 19% increase in the overall G + C content of the 5′ domain. This increase is large by comparison to the 9% evolutionary increase in the total G + C content for 18 S rRNA that has already been found between *Xenopus* and yeast (Salim and Maden, 1981).

The rabbit 5′ sequence, when compared with that of *Xenopus*, reveals

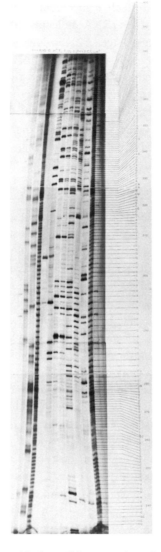

FIGURE 3.   Autoradiogram of both partial enzymatic and chemical digests on 5′-^{32}P-end-labeled 18 S rRNA electrophoresed on a 10% polyacrylamide slab gel in 90% form-amide. Gel dimensions were 0.40 mm thick × 33 cm wide × 140 cm long. Partial digests *(left to right):* (−) minus enzyme; (A > G) chemical digestion; (G) chemical digestion G; (H$^+$) controlled acid hydrolysis; (T$_1$) T$_1$ RNase, 1.25 × 10^{-3} and 1.25 × 10^{-4} U/μg RNA; (U$_2$) U$_2$ RNase, 0.25 × 10^{-4} and 0.25 × 10^{-5} U/μg RNA; (PHY) *Physarum* nuclease I, 0.5 U/10 μg RNA; (B.C.) *Bacillus cereus* RNase, 0.003 U/μg RNA; (CL) chicken liver RNase, 1.0 U/10μg RNA; (H$^+$) controlled acid hydrolysis. Residues 235–400 from the 5′ terminus are indicated.

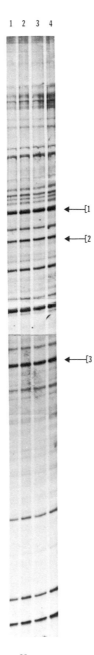

FIGURE 4.   Autoradiogram of the 5'-^{32}P-end-labeled T$_1$ RNase digestion fragments purified on polyacrylamide gel in 7 M urea (80 cm long × 20 cm wide × 0.15 cm thick).

5 -END

```
       10              20              30              40              50
P U A C C U G G U U G A U C C U G C C A G U A G C A U X U G C U U G Y*C U C A A A G A U U A A G C C A
                                            (Am)                (ψ) (ψ)
   51              60              70              80              90              100
U G C(C)U(G)U C U A A G U A C G C A C G G C C G G U A C A G U G A A A C U G(A)G A A U G G C U C A*U
      (A)  (G)                                                      (C)                        (Am)
  101             110             120             130             140             150
U A A A Y*C A G Y*U A U G G U X C C U U X G G U C G C U C G C U C C U C U C C U A C U U G G A U A A
      (ψ)     (ψ)           (Um)        (Um)
 151             160             170             180             190             200
C U G U G G U A X U U C U A G A*G C U A A X A C*A U G C C G(A)C G G(X)G(Q)U G(A)C C C C C U U U G U
              (Am)           (Am)       (Um)(Cm)    (U)      (G) (A)   (U)
 201             210             220             230             240             250
G C G G G A U G C G U G C A U U U A U C A G A U C A A A A C C A A C C C(C)C G U C A G U U U C C C
                                                                    (G)
 251             260             270             280             290             300
C C G G C C C C G C(G)G Y Y Y Y G G U G G G C U C G G C(G)G C U(G)U G G U G A C U C U A G A U A A C
                   (G)                              (C)    (U)
 301             310             320             330             340             350
C U C G G G C C G A U C G C A Y*G C(X)C G U G(G)C G(A)(A)A U U C G A A C G U(C)U G C C C(U)A U C A A
                            (U)      (G)  (A)(A)           (A)         (U)
 351             360             370             380             390             400
C U U U C G A U G G C A G U C G C C G U G U C U A C C A U G G U G A C C A C G G G U G A C G G G G A
```

FIGURE 5.   (A) Nucleotide sequence of the 5′ domain for rabbit 18 S rRNA determined directly from 5′-^{32}P-end-labeled rRNA. (B) Nucleotide sequence of the 3′ domain for rabbit 18 S rRNA determined directly from both 3′-^{32}P-end-labeled rRNA and the 5′-^{32}P-end-labeled T[1] RNase digestion fragments 1, 2, and 3.

a striking 12-nucleotide G-rich insert (261–272) that is also absent in yeast. This insert, as well as 25 of the 44 insertions into this domain, occur within a putative helical stem (Zwieb *et al.*, 1981) thought to exist between residues 222 and 245 of the yeast sequence. Since this region is found to be base-paired using enzymatic structure probes (see below), it is likely that these G + C insertions extend and stabilize a helix already existent within lower eukaryotic rRNA. It is noteworthy that most of the sequences absent from *Xenopus,* but still present in both rabbit and yeast rRNA (240–241, 245–247, 254–257), also reside within this highly mutable region. These apparent deletions in *Xenopus* must have occurred shortly after the divergence of mammals and amphibians from a common ancestor.

In the 5′ domain, the rabbit sequence was found to be 86% homologous with *Xenopus* and 69% homologous with yeast. A more profound homology is evident in the 3′ domain, with 95% homology between rabbit and *Xenopus* and 78% between rabbit and yeast. The greater sequence conservation of the 3′ domain in eukaryotic 18 S rRNA may reflect its likely importance in mRNA and tRNA binding, subunit association, and the function of the 40 S subunit

#1
↓ 801
G A U A

800   790   780   770   760   751
C C G U C G U A G U U C C G A C C A C A A A C G A U G C C G A C U G G C G A U G C G G C G G C G U (C/U)

750   740   730   720   710#2   701
A U U C C C A U G A C C C G A C (G/C) G G (C/A) A (G/C) C U U C C G G G A A A C C A A A G U C U U U G G G U U C

700   690   680   670   660   651
C G G G G G G A G U A U G G U U G C A A A G C (C/U) G A A A C U U A A A G G A A U U G A C G G A A G G G

650   640   630   620   610   601
C A C C A C C A G G A G U G G A (C/U) C C U G C G G C C (U/C) A A U U U G A C X (U/C) A A (U/C) A (U/U) G G G A A A C C
                                                                              (am ψ)

600   590   580   570   560   551
U C A C C C G G C C C G G A C A C G G A C A G G A X U G A C A G A U U G A U A G C U C U U U C U C G
                                          (Um)

550   540   530   520   510   501
A U U C U G U G G (G/C) U G G X G X U G C A U G G C U G U U C U U A G U (C/U)(G/G) U G G A G C G A U U U G U
                        (Um) (Gm)

500   490#3   480   470   460   451
C U G G U U A A U U C C G A U A A C G A X C (G/A) A G A C U X U (G/C) G C A U G C U A A C U A G U C A C G C
                        (Am)   (Cm)

450   440   430   420   410   401
G A C C C C C G A G C G G U C G G C G U C C C C C A A C U X U U C A G A G G D A C A A G U G G C (C/G) U
                                                              (Um)

400   390   380   370   360   351
U C A G C A C C G A G A U U G A G C A A U A A C A X G U C U G U G A (C/U) G C C C U U A G A U G U C C G
                                          (Gm)

350   340   330   320   310   301
G U C G G C A C G C G C G C U A C A C U G A C U G G C U C A G C G U G U G C U (U/C) A C C C U A C G C C

300   290   280   270   260   251
C G G C A G G C C G G (G/U) A A C C C G U U G A A C C C C A U U C G U G A U A G G G A U C G G G G A U U G

250   240   230   220   210   201
C A A U U A U U U C C C A U G (U/A) A C G A G G A A U U C C C A G U A A G U G C G G G U C A U A A G C U
                        (m7G)   (ψ)

200   190   180   170   160   151
U G C G U U G A U U A A G U C C C U G C C C U U U G U A C A C A C C G X C C G U C G C U A C U A C C
          (Am)   (Cm)

150   140   130   120   110   101
G A U U G G A U G G U U U A G U G G G C C C U C G G A U C G G C C C G C C G G G U C G G C C C A C

100   90   80   70   60   51
G G C C C U G G C G G A G C G C U G A G A A G A C G G U C G A A C U X G A C U A U C U A G A G G A A
                                                  (Um)

50   40   30   20   10   1
G U A A A A G U C G U A A G A A G G U U U C C G U A G G U G A ^m_mA^m_mC C U G C G G A A G G A U C A U U A $_{OH}$

3′

FIGURE 5.   Continued

during protein synthesis as has been shown for prokaryotic 16 S rRNA (Noller and Woese, 1981). In recent studies, the 5' nucleotide of the anticodon in *Escherichia coli* N-acetyl-tRNA^{Val} and N-acetyl-tRNA^{Ser} was shown to be covalently cross-linked to prokaryotic 16 S rRNA by UV irradiation when the tRNA occupied the ribosomal P site (Ofengand and Liou, 1980). The specific nucleotide in 16 S rRNA involved in the cross-link was recently identified as C_{1400} (Prince *et al.*, 1982). This C residue lies within a sequence that is also present in eukaryotic 18 S rRNA correponding to nucleotides 160–180 in Fig. 6B. This highly conserved region suggests the likelihood of a similar tRNA–rRNA interaction in eukaryotes. Recent studies indicate that *E. coli* N-acetyl-tRNA^{Val} can indeed be cross-linked to yeast 18 S rRNA when occupying the P site of 40 S subunits (Offengand *et al.*, 1982).

3.2 Secondary Structure Determination

The identification of conserved sequences within rRNA is one method for mapping those regions of the molecule important to its function. To relate primary structure to function, the topography of 18 S rRNA within the 40 S subunit must be determined. Use of both chemical and enzymatic structure probes in solution has recently generated useful information for tRNA (Peattie and Gilbert, 1980), 5 S RNA (Douthwaite and Garrett, 1981), and 16 S rRNA structure (Stiegler *et al.*, 1981b). In this chapter, we report the secondary structure analysis of the 5' and 3' domains for rabbit 18 S rRNA in solution using structure-specific enzymes. In an RNA molecule, the accessibility of a specific region to nucleolytic cleavage is dictated, in part, by its relative exposure to the surface of the molecule and also by the size and structural specificity of the enzyme. Recent studies have shown that both single-strand-specific S_1 nuclease and T_1 RNase can be used successfully for mapping single-stranded regions in various ^{32}P-end-labeled molecules (Wrede *et al.*, 1979). Likewise, base-paired regions within a ^{32}P-end-labeled RNA can be similarly mapped using double-strand-specific ribonuclease V_1 from cobra venom (Favorova *et al.*, 1981; Lockard and Kumar, 1981). This structure-mapping procedure (Wurst *et al.*, 1978) utilizes ^{32}P-end-labeled RNA that is first partially digested with a structure-specific enzyme under nondenaturing conditions, followed by electrophoretic fractionation of the resultant fragments in adjacent lanes of a denaturing polyacrylamide sequencing gel.

Both the 5'- and 3'-^{32}P-end-labeled 18 S rRNAs were first renatured, then digested with the structure-specific enzymes as outlined in Fig. 7. A polyacrylamide gel analysis of a structure map for 3'-^{32}P-end-labeled 18 S rRNA is shown in Fig. 8. Double-strand-specific V_1 scissions are evident along with single-strand-specific S_1 and T_1 nuclease digestions. V_1 nuclease, like S_1 nuclease, generates fragments with 3' hydroxyls and 5' phosphates at the site of cleavage. We find that a complete ladder of fragments with 3'

5'

```
              10              20              30            40          50
XENOPUS  pUACCUGGUUGAUCCUGCCAGUAGCAUAUGCUUGUCUCAAAGAUUAAGCCA
RABBIT   pUACCUGGUUGAUCCUGCCAGUAGCAUAUGCUUGUCUCAAAGAUUAAGCCA
YEAST    pUAUCUGGUUGAUCCUGCCAGUAGCAUAUGCUUGUCUCAAAGAUUAAGCCA
                                     U

             51          60          70          80          90       100
XENOPUS  UGCACGUGUAAGUACGCACGGCCGGUACAGUGAAACUGCGAAUGGCUCAU
RABBIT   UGC(C/A)U(G)UCUAAGUACGCACGGCCGGUACAGUGAAACUG(C/C)GAAUGGCUCAU
YEAST    UGCAUGUCUAAGUAUAUAAGC       UACAGUGAAACUGCGAAUGGCUCAU
                                  AAUUUA

            101         110         120         130         140       150
XENOPUS  UAAAUCAGUUAUGCUUCCUUUGAUCGCUCCAUCUG UU  UACUUGGAUAA
RABBIT   UAAAUCAGUUAUGGUUCCUUUGGUCGCUCGCUCCUCUCCUACUUGGAUAA
YEAST    UAAAUCAGUUAUCGUUUAU     UUGAUAGUCCUUUUACUACAUGGUAUAA
                                                 U

            151         160         170         180    A    190       200
XENOPUS  CUGUGGUAAUUCUAGAGCUAAUACAUGCCGACGGCGCUGACCCCC  A
RABBIT   CUGUGGUAAUUCUAGAGCUAAUACAUGCCG(A)CGG(C)G(G)UG(U)CCCCCUUUGU
YEAST    CCGUGGUAAUUCUAGAGCUAAUACAUGC UUAAAA UCUCACCCUUUGA
                                                         G

            201         210         220         230         240       250
XENOPUS   GGGAUGCGUGCAUUUAUCAGAGCAAAACCAAUCCG G  GG    CCC
RABBIT   GCGGGAUGCGUGCAUUUAUCAGAUCAAAACCAACCC(G)CGUCAGUUUCCCC
YEAST    AGGA G AUGUAUUUAUUAGAUAAAAAUCAAU    GUC   UUC
                                     A

            251         260         270         280         290       300
XENOPUS  CCG    CGC             CCCGGCCGCUUUGGUGACUCUAGAUAAC
RABBIT   CCGGCCCCGC(S)GYYYYGGUGGGCUCGGC(U)GCU(G)UGGUGACUCUAGAUAAC
YEAST     GGACUC              UUUUGAUU  CAUAAUAACUUUUCG AAU
                                            GGC
                                            CG
            301         310      320CC AA AU 330        340       350
XENOPUS  CUCGGGCCGAUCGCACGUCCCGUGCGACAUUCGGAUGUCUGCCCUAUCAA
RABBIT   CUCGGGCCGAUCGCAYGC(C/C)CGUG(C/U)CG(A/C)AUUCGAACGU(C/A)UGCCC(U)AUCAA
YEAST    CGCAGGCC  UUUGU  GCUGGCGAUGGCAUUCAAAUUUCUGCCCUAUCAA
                 U                       UU

            351         360         370         380         390       400
XENOPUS  CUUUCGAUGGUACUUUCUGCGCCUACCAUGGUGACCACGGGUAACGGGGA
RABBIT   CUUUCGAUGGCAGUCGCCGUGUCUACCAUGGUGACCACGGGUGACGGGGA
YEAST    CUUUCGAUGGUAGGAUAGUGGCCUACCAUGGUUUCAACGGUAAUACGGG
```

FIGURE 6A. Comparison of the rabbit nucleotide sequence at the 5' domain with that from *X. laevis* and yeast 18 S rRNA.

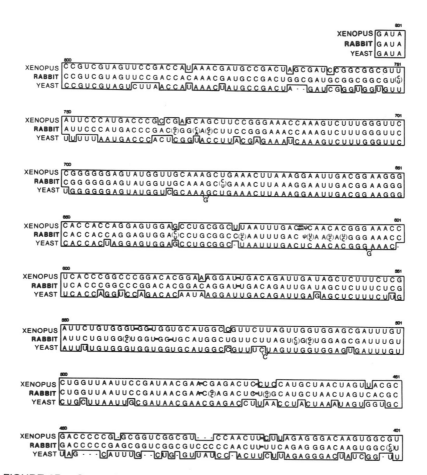

FIGURE 6B.   Comparison of the rabbit nucleotide sequence at the 3′ domain with that from *X. laevis* and yeast 18 S rRNA.

hydroxyls can easily be generated on partial digestion of ^{32}P-end-labeled RNA with *Neurospora crassa* endonuclease (N.E.). When loaded onto a lane between both S_1 and V_1 partial digests as shown in Fig. 8, the *N. crassa* digest facilitates simultaneous interpretation of both single- and double-strand-specific cleavages.

Structure maps of the 5′ proximal 400 residues and the 3′ distal 300 nucleotides for rabbit 18 S rRNA are shown in Fig. 9. Alternating cleavages with double- and single-strand-specific structure probes are evident. In a few regions, there is cleavage with both S_1 and V_1 nucleases. In these areas, it

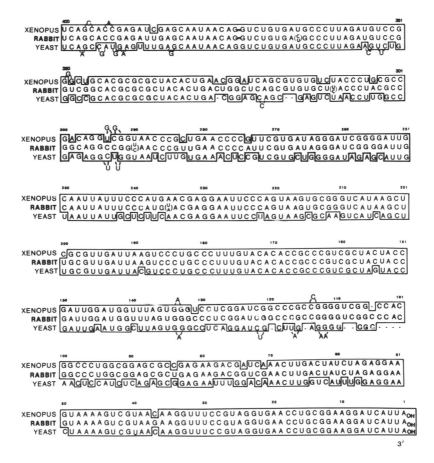

FIGURE 6B. *Continued*

is not clear whether such simultaneous cleavages are due to secondary cutting with one or both enzymes or to recognition of tertiary interactions by V_1 nuclease. The amount of structural detail generated with these enzymatic probes is significant. Since we have mapped the structure of only approximately 700 nucleotides thus far, we have not yet generated computer-aided structure predictions for these relatively short regions. However, a general secondary-structure model for prokaryotic small rRNA has recently been proposed. This model is based on enzymatic and chemical structure-probing, RNA–RNA cross-linking, thermodynamic folding, and phylogenetic se-

RENATURATION OF NATIVE 18SrRNA

1. PREINCUBATION IN STRUCTURE BUFFER FOR 10 MIN.
    at 70°C, THEN SLOW COOLED.

OR

2. PREINCUBATION IN 20mM TRIS-HCl pH 7.2 FOR 5 MIN. AT 85° C, THEN
    SLOW COOLED IN STRUCTURE BUFFER.

DIGESTION OF 18SrRNA

SINGLE-STRAND SPECIFIC

   S, NUCLEASE:
   $$\begin{array}{ll} \text{structure} \\ \text{buffer} \end{array}\left[\begin{array}{ll} 40\text{mM} & \text{NaOAc, pH 4.5} \\ 5\text{mM} & \text{MgCl}_2 \\ 50\text{mM} & \text{KCl} \\ 1\text{mM} & \text{ZnSO}_4 \end{array}\right.$$

   0.003 UNITS/μgRNA AT 37°C,
   TIME POINTS REMOVED AT 1′, 5′, 10′, 20′

   T, NUCLEASE:
   $$\begin{array}{ll} \text{structure} \\ \text{buffer} \end{array}\left[\begin{array}{ll} 15\text{mM} & \text{NaOAc, pH 5.0} \\ 15\text{mM} & \text{MgCl}_2 \\ 300\text{mM} & \text{NaCl} \end{array}\right.$$

   1 $\times$ 10^{-4} UNITS/μgRNA AT 37°C fpr 5 MIN.

DOUBLE-STRAND SPECIFIC

   V, NUCLEASE:
   $$\begin{array}{ll} \text{structure} \\ \text{buffer} \end{array}\left[\begin{array}{ll} 20\text{mM} & \text{TRIS-HCl pH 7.2} \\ 10\text{mM} & \text{MgCl}_2 \\ 200\text{mM} & \text{NaCl} \end{array}\right.$$

   1 $\times$ 10^{-8} UNITS/μgRNA AT 37°C,
   TIME POINTS REMOVED AT 1′, 5′, 10′, 20′

FIGURE 7. Conditions for enzymatic digestion of ^{32}P-end-labeled 18 S rRNA using single- and double-strand-specific structure probes.

FIGURE 8. Autoradiogram of partial digests on 3′-^{32}P-end-labeled 18 S rRNA electrophoresed on a 10% polyacrylamide slab gel in 90% formamide. Gel dimensions were 0.40 mm thick $\times$ 33 cm wide $\times$ 140 cm long. *Left to right:* ($-$) minus enzyme; (V$_1$) V$_1$ nuclease, 20 min, 10 min, 5 min, 1 min; (N.E.) *N. crassa* endonuclease, 5 $\times$ 10^{-2} U/μg RNA; (S$_1$) S$_1$ nuclease, 1 min, 5 min, 10 min, 20 min; (T$_1$S) T$_1$ RNase under nondenaturing conditions, 3 $\times$ 10^{-4} and 3 $\times$ 10^{-5} U/μg RNA; (H$^+$) controlled acid hydrolysis; (T$_1$) T$_1$ RNase under denaturing conditions, 5 $\times$ 10^{-4} and 5 $\times$ 10^{-3} U/μg RNA. Residues 120–239 from the 3′ terminus are indicated.

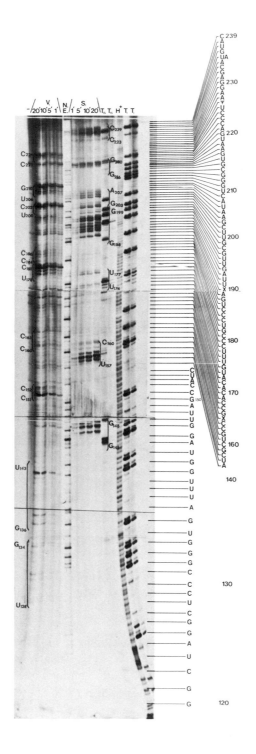

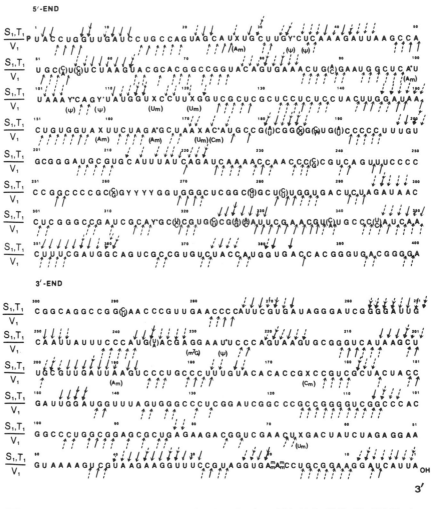

FIGURE 9.   Structure map of the 5' and 3' domains for rabbit 18 S rRNA. (S_1, T_1) Single-strand-specific cutting; (V_1) double-strand-specific cutting.

quence comparisons within various prokaryotic and organelle small rRNAs (Noller and Woese, 1981; Zwieb *et al.*, 1981; Stiegler *et al.*, 1981b). In summary, the prokaryotic small rRNAs can be folded into a common basic structure, regardless of their origins. Specific regions of the molecule appear to be highly conserved evolutionarily, in terms of both primary and secondary structure. These regions are most likely involved in the maintenance of a fundamental conformation. Although very few structure data are available for eukaryotic small rRNA, both the yeast and the *Xenopus* nucleotide sequences can be folded into a structure maintaining many of the prominent topographical features evident in prokaryotic 16 S rRNA (Zwieb *et al.*, 1981; Stiegler *et al.*, 1981a,b). This is remarkable considering both the differences in length and the low sequence homology between prokaryotic and eukaryotic small rRNA (Rubstov *et al.*, 1980; Salim and Maden, 1981). Interestingly, the nucleotide sequences surrounding the G residues cleaved with T_1 RNase in high-salt buffer generating the rabbit fragments 1, 2, and 3 not only are highly conserved in eukaryotic 18 S rRNA as shown in Fig. 6B, but also are conserved within *E. coli* 16 S rRNA and reside within single-stranded regions (Stiegler *et al.*, 1981a,b; Noller and Woese, 1981). These same G residues are reactive to kethoxal and are cleaved with T_1 RNase in *E. coli* 30 S subunits (Noller and Woese, 1981), underscoring their likely importance in the structure and function of the small ribosomal subunit. However, there are also many regions within the structure map of the rabbit 5′ and 3′ domains shown in Fig. 9 that do not agree with the secondary structure proposed for eukaryotic 18 S rRNA (Stiegler *et al.*, 1981a,b; Zwieb *et al.*, 1981). Obviously, further refinement of the eukaryotic 18 S rRNA structure model is necessary and will require not only additional rRNA sequences, but also a substantial amount of enzymatic and chemical structure data of the type indicated here.

4. DISCUSSION

To begin to understand the functional organization of the eukaryotic 40 S ribosomal subunit with respect to initiation factor association, mRNA and rRNA binding, and subunit joining, both the primary and secondary structure of its 18 S rRNA component must be determined. The methods described here for nucleotide-sequence analysis of rabbit 18 S directly are general and can be used for sequence analysis of most comparably large and highly structured RNA molecules.

We are currently completing the nucleotide sequence using additional fragments generated by site-directed cleavage with RNase H and sequence-specific DNA oligomers. Sequence analysis of 18 S rRNA directly allows for the mapping of most of the modified bases, as well as detection of apparent cistron heterogeneities.

The methods for ^{32}P-end-labeling large RNA molecules to high specific activity, use of thin 140-cm-long formamide polyacrylamide sequencing gels, and particularly cleavage of 18 S rRNA into defined fragments now permit direct secondary-structure determination of mammalian 18 S rRNA using structure-specific enzymatic and chemical probes. The establishment of a fundamental conformation for mammalian 18 S rRNA within the 40 S subunit will be imperative in eventually defining the specific functions of its various domains.

ACKNOWLEDGMENTS.   The work reported herein was supported by Grant GM 28985 from the National Institutes of Health (R. E. L.) and National Science Foundation Grant PCM 82 10176 (A. K.). For J. F. C. this work is in partial fulfillment of the Ph.D. requirements at The George Washington University.

REFERENCES

Choi, Y. C., and Busch, H., 1978, Modified nucleotides in T$_1$ RNase oligonucleotides of 18 S ribosomal RNA of the Novikoff hepatoma, *Biochemistry* **17**:2551–2560.

Donis-Keller, H., Maxam, A. M., and Gilbert, W., 1977, Mapping adenines, guanines and pyrimidines in RNA, *Nucleic Acids Res.* **4**:2527–2538.

Douthwaite, S., and Garrett, R. A., 1981, Secondary structure of prokaryotic 5 S ribosomal ribonucleic acids: A study with ribonucleases, *Biochemistry* **20**:7301–7307.

Favorova, O. O., Fasiolo, F., Keith, G., Vassilenko, S. K., and Ebel, J.-P., 1981, Partial digestion of tRNA aminoacyl–tRNA synthetase complexes with cobra venom ribonuclease, *Biochemistry* **20**:1006–1011.

Lake, J. A., 1980, Ribosome structure and functional sites, in: *Ribosomes: Structure, Function and Genetics* (G. Chambliss, G. R. Craven, J. Davies, K. Davis, L. Kahan, and M. Nomura, eds.), pp. 207–236, University Park Press, Baltimore.

Lockard, R. E., and Kumar, A., 1981, Mapping tRNA structure in solution using double-strand specific ribonuclease V$_1$ from cobra venom, *Nucleic Acids Res.* **9**:5125–5140.

Lockard, R. E., and RajBhandary, U. L., 1976, Nucleotide sequence at the 5′ termini of rabbit α and β globing mRNA, *Cell* **9**:747–760.

Lockard, R. E., Alzner-Deweerd, B., Heckman, J. E., MacGee, J., Tabor, M. W., and RajBhandary, U. L., 1978, Sequence analysis of 5′-^{32}P-labeled mRNA and tRNA using polyacrylamide gel electrophoresis, *Nucleic Acids Res.* **5**:37–56.

Lockard, R. E., Connaughton, J. F., and Kumar, A., 1982, Nucleotide sequence of the 5′ and 3′ domains for rabbit 18 S ribosomal RNA, *Nucleic Acids Res.* **10**:3445–3457.

Maden, B. E. H., and Kahn, M. S. N., 1977, Methylated nucleotide sequences in HeLa-cell ribosomal ribonucleic acid, *Biochem. J.* **167**:211–221.

Maden, B. E. H., and Salim, M., 1974, The methylated nucleotide sequences in HeLa cell ribosomal RNA and its precursors, *J. Mol. Biol.* **88**:133–164.

Noller, H. F., and Woese, C. R., 1981, Secondary structure of 16 S ribosomal RNA, *Science* **212**:403–411.

Ofengand, J., and Liou, R., 1980, Evidence for pyrimidine–pyrimidine cyclobutane dimer formation in the covalent cross-linking between transfer ribonucleic acid and 16 S ribonucleic acid at the ribosomal P site, *Biochemistry* **18:**4814–4822.

Ofengand, J., Gornicki, P., Chakraburtty, K., and Nurse, K., 1982, Functional conservation near the 3′ end of eukaryotic small subunit RNA: Photochemical cross-linking of P site-bound acetylvalyl-tRNA to 18 S RNA of yeast ribosomes, *Proc. Natl. Acad. Sci. U.S.A.* **79:**2807–2812.

Peattie, D. A., 1979, Direct chemical method for sequencing RNA, *Proc. Natl. Acad. Sci. U.S.A.* **76:**1760–1764.

Peattie, D. A., and Gilbert, W., 1980, Chemical probes for higher-order structure in RNA, *Proc. Natl. Acad. Sci. U.S.A.* **77:**4679–4682.

Prince, J. B., Taylor, B. H., Thurlow, D. L., Ofengand, J., and Zimmerman, R. A., 1982, Covalent crosslinking of tRNA^{Val} to 16S RNA at the ribosomal P site: Identification of crosslinked residues, *Proc. Natl. Acad. Sci. U.S.A.* **79:**5450–5454.

Rubstov, P. M., Musakhanov, M. M., Zakharyev, V. M., Krayev, A. S., Skryabin, K. G., and Bayer, A. A., 1980, The structure of the yeast ribosomal RNA genes. I. The complete nucleotide sequence of the 18S ribosomal RNA gene from *Saccharomyces cerevisiae*, *Nucleic Acids Res.* **8:**5779–5794.

Salim, M., and Maden, B. E. H., 1981, Nucleotide sequence of *Xenopus laevis* 18S ribosomal RNA inferred from gene sequence, *Nature (London)* **291:**205–208.

Stiegler, P., Carbon, P., Ebel, J.-P., and Ehresmann, C., 1981a, A general secondary structure model for procaryotic and eucaryotic RNAs of the small ribosomal subunits, *Eur. J. Biochem.* **120:**487–495.

Stiegler, P., Carbon, P., Zuker, M., Ebel, J.-P., and Ehresmann, C., 1981b, Structural organization of the 16S RNA from *E. coli:* Topography and secondary structure, *Nucleic Acids Res.* **9:**2153–2206.

Wrede, P., Wurst, R., Vournakis, J., and Rich, A., 1979, Conformational changes of yeast tRNA^{Phe} and *E. coli* tRNA$_2^{Glu}$ as indicated by different nuclease digestion patterns, *J. Biol. Chem.* **254:**9068–9615.

Wurst, R. M., Vournakis, J. J., and Maxam, A. M., 1978, Structure mapping of 5′-^{32}P-labeled RNA with S$_1$ nuclease, *Biochemistry* **17:**4493–4499.

Zwieb, C., Glotz, C., and Brimaccombe, R., 1981, Secondary structure comparisons between small subunit ribosomal RNA molecules from six different species, *Nucleic Acids Res.* **9:**3621–3640.

Index